Christian Rudder is President and [illegible] author of the popular stats blog O[illegible]Trends. He graduated from Harvard in 1998 with a degree in maths. His work has been written about in the *New York Times* and the *New Yorker*, among other places. He lives in Brooklyn with his wife and daughter.

Praise for *Dataclysm*:

'A wonderful march through infographics created using data derived from the web. ... A fun, visual book – and a necessary one at that' Max Wallis, *Independent*, Books of the Year

'At a time when consumers are increasingly wary of online tracking, Rudder makes a powerful argument in *Dataclysm* that the ability to tell so much about us from the trails we leave is as potentially useful as it is pernicious, and as educational as it may be unsettling' *Financial Times*

'There's another side of Big Data you haven't seen – not the one that promised to use our digital world to our advantage to optimize, monetize, or systematize every last part our lives. It's the Big Data that rears its ugly head and tells us what we *don't* want to know. And that, as Christian Rudder demonstrates in his new book, *Dataclysm*, is perhaps an equally worthwhile pursuit. Before we heighten the human experience, we should understand it first' *TIME*

'Studying human behavior is a little like exploring a jungle: It's messy, hard, and easy to lose your way. But Christian Rudder is a consummate guide, revealing essential truths about who we are. Big Data has never been so fun' Dan Ariely, author of *Predictably Irrational*

'Rudder draws from big data sets – Google searches, Twitter updates, illicitly obtained Facebook data passed shiftily between researchers like bags of weed – to draw out subtle patterns in politics, sexuality, identity and behavior that are only revealed with distance and aggregation. … *Dataclysm* will entertain those who want to know how machines see us. It also serves as a call to action, showing us how server farms running everything from home shopping to homeland security turn us into easily digested data products. Rudder's message is clear: In this particular sausage factory, we are the pigs' *New Scientist*

'As a researcher, Mr Rudder clearly possesses the statistical acumen to answer the questions he has posed so well. As a writer, he keeps the book moving while fully exploring each topic, revealing his graphs and charts with both explanatory and narrative skill. Though he forgoes statistical particulars like p-values and confidence intervals, he gives an approachable, persuasive account of his data sources and results. He offers explanations of what the data can and cannot tell us, why it is sufficient or insufficient to answer some question we may have, and, if the latter is the case, what sufficient data would look like. He shows you, in short, how to think about data' *Wall Street Journal*

'Rudder is the cofounder of the dating site OkCupid and the data scientist behind its now-legendary trend analyses, but he is also – as it becomes immediately clear from his elegant writing and wildly cross-disciplinary references – a lover of literature, philosophy, anthropology, and all the other humanities that make us human and that, importantly in this case, enhance and ennoble the hard data with dimensional insight into the richness of the human experience … an extraordinarily unusual and dimensional lens on what Carl Sagan memorably called "the aggregate of our joy and suffering"'

Maria Popova, *Brain Pickings*

'*Dataclysm* is a book full of juicy secrets – secrets about who we love, what we crave, why we like, and how we change each other's minds and lives, often without even knowing it. Christian Rudder makes this mathematical narrative of our culture fun to read and even more fun to discuss: You will find yourself sharing these intriguing data-driven revelations with everyone you know'

Jane McGonigal, author of *Reality Is Broken*

'*Dataclysm* is a well-written and funny look at what the numbers reveal about human behavior in the age of social media. It's both profound and a bit disturbing, because, sad to say, we're generally not the kind of people we like to think – or say – we are' *Salon*

'[Rudder] is a quant with soul, and we're lucky to have him' *Elle*

'For all its data and its seemingly dating-specific focus, *Dataclysm* tells the story set forth by the book's subtitle in an entertaining and accessible way. Informative, eye-opening, and (gasp) fun to read. Even if you're not a giant stat head' *Grantland*

'Most data-hyping books are vapor and slogans. This one has the real stuff: actual data and actual analysis taking place on the page. That's something to be praised, loudly and at length. Praiseworthy, too, is Rudder's writing, which is consistently zingy and mercifully free of Silicon Valley business gabble' Jordan Ellenberg, *Washington Post*

'In the first few pages of *Dataclysm*, Christian Rudder uses massive amounts of actual behavioral data to prove what I always believed in my heart: Belle and Sebastian is the whitest band ever. It only gets better from there' Aziz Ansari

'[Rudder] doesn't wring or clap his hands over the Big Data phenomenon (see NSA, Google ads, that sneaky Fitbit) so much as plunge them into Big Data and attempt to pull strange creatures from the murky depths' *New Yorker*

'It's unheard of for a book about Big Data to read like a guilty pleasure, but *Dataclysm* does. It's a fascinating, almost voyeuristic look at who we really are and what we really want'

Steven Strogatz, Schurman Professor of Applied Mathematics, Cornell University, author of *The Joy of x*

'A hopeful and exciting journey into the heart of data collection ... [Rudder's] book delivers both insider access and a savvy critique of the very machinery he is employed by. Since he's been in the data mines and has risen above them, Rudder becomes a singular and trustworthy guide' *Globe and Mail*

'Compulsively readable – including for those with no particular affinity for numbers in and of themselves – and surprisingly personal. Starting with aggregates, Rudder posits, we can zoom in on the details of how we live, love, fight, work, play, and age; from numbers, we can derive narrative. There are few characters in the book, and few anecdotes – but the human story resounds throughout' *Refinery29*

'Smart, revealing, and sometimes sobering, *Dataclysm* affirms what we probably suspected in our darker moments: When it comes to romance, what we say we want isn't what will actually make us happy. Christian Rudder has tapped the tremendous wealth of data that the Internet offers to tease out thoughts on topics like beauty and race that most of us wouldn't cop to publicly. It's a riveting read, and Rudder is an affable and humane guide'

Adelle Waldman, author of *The Love Affairs of Nathaniel P.*

"Rudder's lively, clear prose ... makes heady concepts understandable and transforms the book's many charts into revealing truths. ... Rudder teaches us a bit about how wonderfully peculiar humans are, and how we go about hiding it' *Flavorwire*

'*Dataclysm* is all about what we can learn about human minds and hearts by analyzing the massive ongoing experiment that is the Internet' *Forbes*

'The book reads as if it's written (well) by a curious child whose parents beg him or her to stop asking "what if" questions. Rudder examines the data of the website he helped create with unwavering curiosity. Every turn presents new questions to be answered, and he happily heads down the rabbit hole to resolve them'

U.S. News

'Christian Rudder has written a funny and profound book about important issues. Race, love, sex – you name it. Are we the sum of the data we produce? Read this book immediately and see if you can answer the question' Errol Morris

'*Dataclysm* offers both the satisfaction of confirming stereotypes and the fun of defying them. ... Such candor is disarming, as is Mr Rudder's puckish sense of humor' *Pittsburgh Post-Gazette*

'Big Data can be like a 3-D movie without 3-D glasses – you know there's a lot going on but you're mainly just disoriented. We should feel fortunate to have an interpreter as skilled (and funny) as Christian Rudder. *Dataclysm* is filled with insights that boil down Big Data into byte-sized revelations' Michael Norton, Harvard Business School, coauthor of *Happy Money*

'With a zest for both the profound and the wacky, Rudder demonstrates how the information we provide individually tells a vast deal about who we are collectively. A visually engaging read and a fascinating topic make this a great choice not just for followers of Nate Silver and fans of infographics, but for just about anyone who, by participating in online activity, has contributed to the data set' *Library Journal*

'Demographers, entrepreneurs, students of history and sociology, and ordinary citizens alike will find plenty of provocations and, yes, much data in Rudder's well-argued, revealing pages'

Kirkus Reviews

Dataclysm

CONTENTS

Part 2.

What Pulls Us Apart

Part 3.

What Makes Us Who We Are

Introduction

You have by now heard a lot about Big Data: the vast potential, the ominous consequences, the paradigm-destroying *new paradigm* it portends for mankind and his ever-loving websites. The mind reels, as if struck by a very dull object. So I don't come here with more hype or reportage on the data phenomenon. I come with the thing itself: the data, phenomenon stripped away. I come with a large store of the actual information that's being collected, which luck, work, wheedling, and more luck have put me in the unique position to possess and analyze.

I was one of the founders of OkCupid, a dating website that, over a very un-bubbly long haul of ten years, has become one of the largest in the world. I started it with three friends. We were all mathematically minded, and the site succeeded in large part because we applied that mind-set to dating; we brought some analysis and rigor to what had historically been the domain of love "experts" and warlocks like Dr. Phil. How the site works isn't all that sophisticated—it turns out the only math you need to model the process of two people getting to know each other is some sober arithmetic—but for whatever reason, our approach resonated, and this year alone 10 million people will use the site to find someone.

As I know too well, websites (and founders of websites) love to throw out big numbers, and most thinking people have no doubt learned to ignore them; you hear millions of this and billions of that and know it's basically "Hooray for me," said with trailing zeros. Unlike Google, Facebook, Twitter, and the other sources whose data will figure prominently in this book, OkCupid is far from a household name—if you and your friends have all been happily married for years, you've probably never heard of us. So I've thought a lot about how to describe the reach of the site to someone who's never used it and who rightly doesn't care about the user-engagement metrics of some guy's startup. I'll put it in personal terms instead. Tonight, some thirty thousand couples will have their first date because of OkCupid. Roughly three thousand of them will end up together long-term. Two hundred of those will get married, and many of them, of course, will have kids. There are children alive and pouting today, grouchy little humans refusing to put their shoes on *right now,* who would never have existed but for the whims of our HTML.

I have no smug idea that we've perfected anything, and it's worth saying here that while I'm proud of the site my friends and I started, I honestly don't care if you're a member or go create an account or what. I've never been on an online date in my life and neither have any of the other founders, and if it's not for you, believe me, I get that. Tech evangelism is one of my least favorite things, and I'm not here to trade my blinking digital beads for anyone's precious island. I still subscribe to magazines. I get the *Times* on the weekend. Tweeting embarrasses me. I can't convince you to use, respect, or "believe in" the Internet or social media any more than you already do—or don't. By all means, keep right on thinking what you've been thinking about the online universe. But if there's one thing I sincerely hope this book might get you to reconsider, it's what you think about yourself. Because that's what this book is really about. OkCupid is just how I arrived at the story.

I have led OkCupid's analytics team since 2009, and my job is to make sense of the data our users create. While my three founding partners have done almost all the hard work of actually building the site, I've spent years just playing with the numbers. Some of what I work on helps us run the business: for example, understanding how men and women view sex and beauty differently is essential for a dating site. But a lot of my results aren't directly useful—just interesting. There's not much you can do with the fact that, statistically, the least black band on Earth is Belle & Sebastian, or that the flash in a snapshot makes a person look seven years older, except to say *huh,* and maybe repeat it at a dinner party. That's basically all we did with this stuff for a while; the insights we gleaned went no further than an occasional lame press release. But eventually we were analyzing enough information that larger trends became apparent, big patterns in the small ones, and, even better, I realized I could use the data to examine taboos like race by direct inspection. That is, instead of asking people survey questions or contriving small-scale experiments, which was how social science was often done in the past, I could go and look at *what actually happens* when, say, 100,000 white men and 100,000 black women interact in private. The data was sitting right there on our servers. It was an irresistible sociological opportunity.

I dug in, and as discoveries built up, like anyone with more ideas than audience, I started a blog to share them with the world. That blog then became this book, after one important improvement. For *Dataclysm,* I've gone far beyond OkCupid. In fact, I've probably put together a data set of person-to-person interaction that's deeper and more varied than anything held by any other private individual—spanning most, if not all, of the significant online data sources of our time. In these pages I'll use my data to speak not just to the habits of one site's users but also to a set of universals.

The public discussion of data has focused primarily on two things: government spying and commercial opportunity. About the first, I doubt I know any more than you—only what I've read. To my knowledge, the national security apparatus has never approached any dating site for access, and unless they plan to criminalize the faceless display of utterly ripped abs or young women from Brooklyn going on and on about how much they like scotch, when, come on, you know they really don't, I can't imagine they'd find much of interest. About the second story, data-as-dollars, I know better. As I was beginning this book, the tech press was slick with drool over the Facebook IPO; they'd collected everyone's personal data and had been turning it into all this money, and now they were about to turn *that* money into even more money in the public markets. A *Times* headline from three days before the offering says it all: "Facebook Must Spin Data into Gold." You half expected Rumpelstiltskin to show up on the OpEd page and be like, "Yes, America, this is a solid buy."

As a founder of an ad-supported site, I can confirm that data *is* useful for selling. Each page of a website can absorb a user's entire experience—everything he clicks, whatever he types, even how long he lingers—and from this it's not hard to form a clear picture of his appetites and how to sate them. But awesome though the power may be, I'm not here to go over our nation's occult mission to sell body spray to people who update their friends about body spray. Given the same access to the data, I am going to put that user experience—the clicks, keystrokes, and milliseconds—to another end. If Big Data's two running stories have been surveillance and money, for the last three years I've been working on a third: the human story.

Facebook might know that you're one of M&M's many fans and send you offers accordingly. They also know when you break up with your boyfriend, move to Texas, begin appearing in lots of pictures with your ex, and start dating him again. Google knows when you're looking for a new car and can show the make and model preselected for just your psychographic. A thrill-seeking socially conscious Type B, M, 25–34? Here's your Subaru. At the same time, Google also knows if you're gay or angry or lonely or racist or worried that your mom has cancer. Twitter, Reddit, Tumblr, Instagram, all these companies are businesses first, but, as a close second, they're demographers of unprecedented reach, thoroughness, and importance. Practically as an accident, digital data can now show us how we fight, how we love, how we age, who we are, and how we're changing. All we have to do is look: from just a very slight remove, the data reveals how people behave when they think no one is watching. Here I will show you what I've seen. Also, fuck body spray.

∞

If you read a lot of popular nonfiction, there is something in *Dataclysm* that you might find unusual. This is a book that deals in aggregates and big numbers, and that makes for a curious absence in a story supposedly about people: there are very few individuals here. Graphs and charts and tables appear in abundance, but there are almost no names. It's become a cliché of pop science to use something small and quirky as a lens for big events—to tell the history of the world via a turnip, to trace a war back to a fish, to shine a penlight through a prism *just so* and cast the whole pretty rainbow on your bedroom wall. I'm going in the opposite direction. I'm taking something big—an enormous set of what people are doing and thinking and saying, terabytes of data—and filtering from it many small things: what your network of friends says about the stability of your marriage, how Asians (and whites and blacks and Latinos) are least likely to describe themselves, where and why gay people stay in the closet, how writing has changed in the last ten years, and how anger hasn't. The idea is to move our understanding of ourselves away from narratives and toward numbers, or, rather, to think in such a way that numbers *are* the narrative.

This approach evolved from long toil in the statistical slag pits. *Dataclysm* is an extension of what my coworkers and I have been doing for years. A dating site brings people together, and to do that credibly it has to get at their desires, habits, and revulsions. So you collect a lot of detailed data and work very hard to translate it all into general theories of human behavior. What a person develops working amidst all this information, as opposed to, say, working for the wedding section of the Sunday paper, is a special kinship with the shambling whole of humanity rather than with any two individuals. You grow to understand people much as a chemist might understand, and through understanding come to love, the swirling molecules of his tincture.

That said, all websites, and indeed all data scientists, objectify. Algorithms don't work well with things that aren't numbers, so when you want a computer to understand an idea, you have to convert as much of it as you can into digits. The challenge facing sites and apps is thus to chop and jam the continuum of human experience into little buckets 1, 2, 3, without anyone noticing: to divide some vast, ineffable process—for Facebook, friendship, for Reddit, community, for dating sites, love—into pieces a server can handle. At the same time you have to retain as much of the je ne sais quoi of the thing as you can, so the users believe what you're offering represents real life. It's a delicate illusion, the Internet; imagine a carrot sliced so cleanly that the pieces stay there in place on the cutting board, still in the shape of a carrot. And while this tension—between the continuity of the human condition and the fracture of the database—can make running a website complicated, it's also what makes my story go. The approximations technology has devised for things like lust and friendship offer a truly novel opportunity: to put hard numbers to some timeless mysteries; to take experiences that we've been content to put aside as "unquantifiable" and instead gain some understanding. As the approximations have gotten better and better, and as people have allowed them further into their lives, that understanding has improved with startling speed. I'm going to give you a quick example, but I first want to say that "Making the Ineffable Totally Effable" really should've been OkCupid's tagline. Alas.

Ratings are everywhere on the Internet. Whether it's Reddit's up/down votes, Amazon's customer reviews, or even Facebook's "like" button, websites ask you to vote because that vote turns something fluid and idiosyncratic—your opinion—into something they can understand and use. Dating sites ask people to rate one another because it lets them transform first impressions such as:

> He's got beautiful eyes
> Hmmm, he's cute, but I don't like redheads
> Ugh, gross

... into simple numbers, say, 5, 3, 1 on a five-star scale. Sites have collected billions of these microjudgments, one person's snap opinion of someone else. Together, all those tiny thoughts form a source of vast insight into how people arrive at opinions of one another.

The most basic thing you can do with person-to-person ratings like this is count them up. Take a census of how many people averaged one star, two stars, and so on, and then compare the tallies. At right, I've done just that with the average votes given to straight women by straight men. This is the shape of the curve:

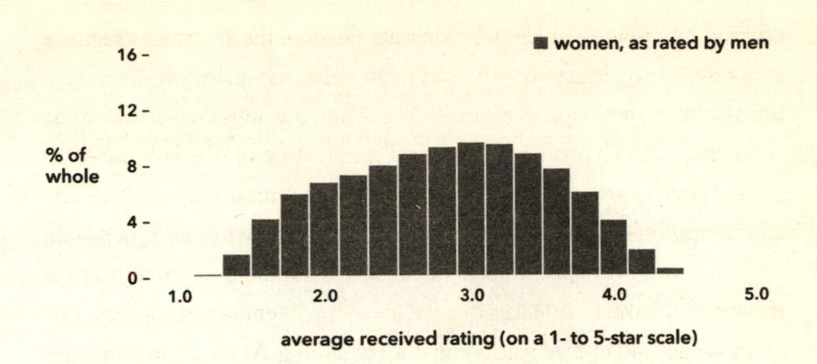

Fifty-one million preferences boil down to this simple stand of rectangles. It is, in essence, the collected male opinion of female beauty on OkCupid. It folds all the tiny stories (what a man thinks of a woman, millions of times over) and all the anecdotes (any one of which we could've expanded upon, were this a different kind of book) into an intelligible whole. Looking at people like this is like looking at Earth from space; you lose the detail, but you get to see something familiar in a totally new way.

So what is this curve telling us? It's easy to take this basic shape—a bell curve—for granted, because examples in textbooks have probably led you to expect it, but the scores could easily have gone hard to one side or the other. When personal preference is involved, they often do. Take ratings of pizza joints on Foursquare, which tend to be very positive:

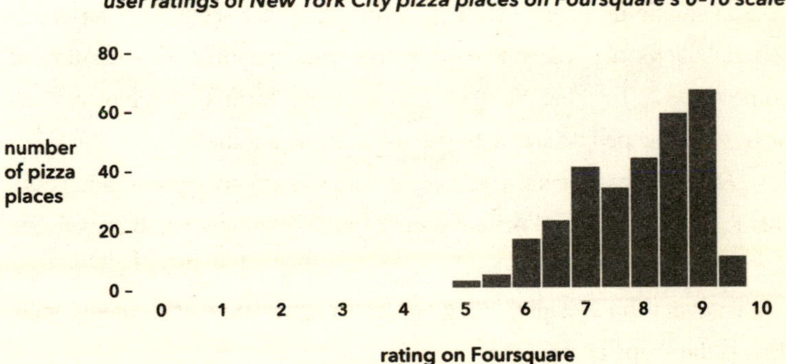

Or take the recent approval ratings for Congress, which, because politicians are the moral opposite of pizza, skew the other way:

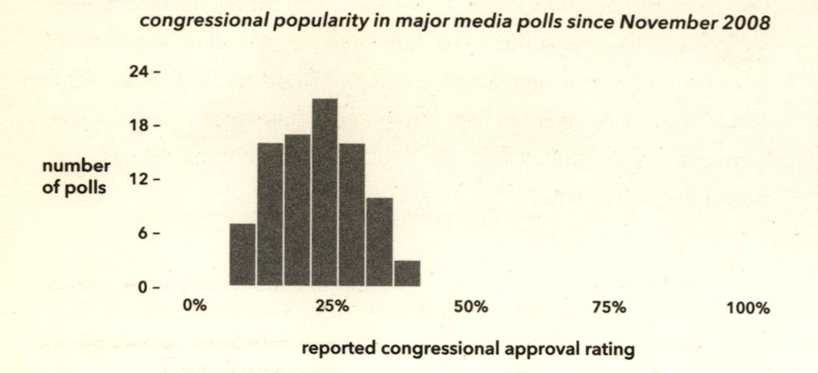

Also, our male-to-female ratings curve is *unimodal,* meaning that the women's scores tend to cluster around a single value. This again is easy to shrug at, but many situations have multiple modes, or "typical" values. If you plot NBA players by how often they were in the starting lineup in the 2012–13 season, you get a bunch of athletes clustered at either end, and almost no one in the middle:

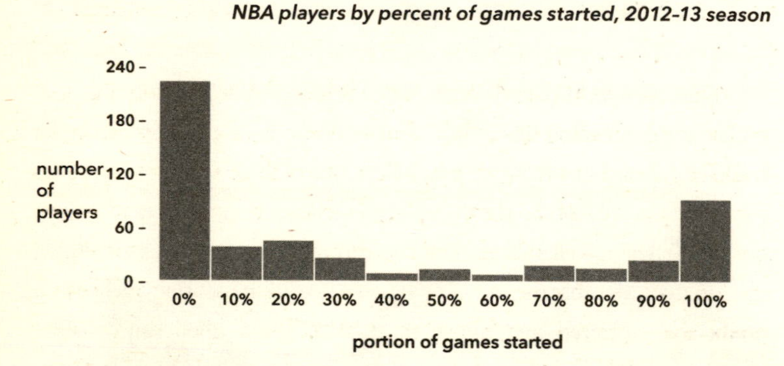

That's the data telling us that coaches think a given player is either good enough to start, or he isn't, and the guy's in or out of the lineup accordingly. There's a clear binary system. Similarly, in our ratings data, men as a group might've seen women as "gorgeous" or "ugly" and left it at that; like top-line basketball talent, beauty could've been a you-have-it-or-you-don't kind of thing. But the curve we started with says something else. Looking for understanding in data is often a matter of considering your results against these kinds of counterfactuals. Sometimes, in the face of an infinity of alternatives, a straightforward result is all the more remarkable for being so. In fact, our graph is quite close to what's called a *symmetric beta distribution*—a curve often deployed to model basic unbiased decisions—which I'll overlay here:

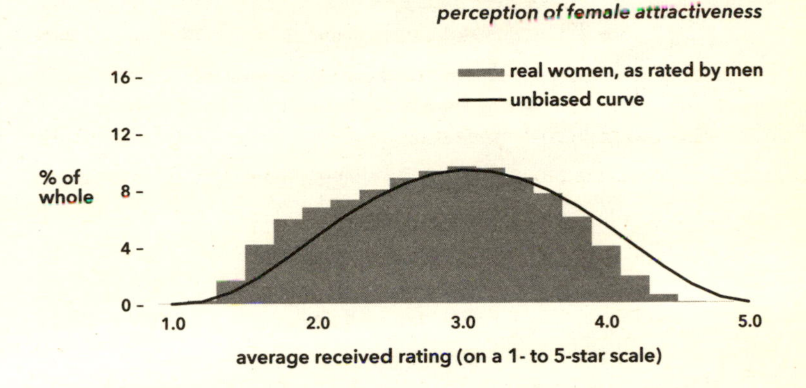

Our real-world data diverges only slightly (6 percent) from this formulaic ideal, meaning this graph of male desire is more or less what we could've guessed in a vacuum: it is, in fact, one of those textbook examples I was making light of. So the curve is predictable, centered—maybe even boring. So what? Well, this is a rare context where boringness is something special: it implies that the individual men who did the scoring are likewise predictable, centered, and, above all, unbiased. And when you consider the supermodels, the porn, the cover girls, the Lara Croft–style fembots, the Bud Light ads, and, most devious of all, the Photoshop jobs that surely these men see every day, the fact that male opinion of female attractiveness is still where it's supposed to be is, by my lights, a small miracle. It's practically common sense that men should have unrealistic expectations of women's looks, and yet here we see it's just not true. In any event, they're far more generous than the women, whose votes go like this:

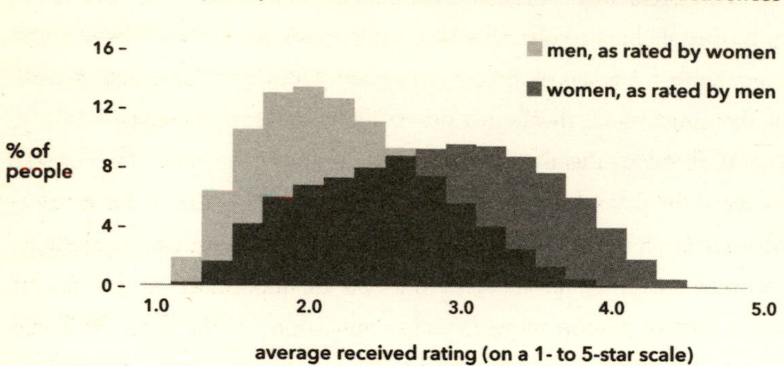

The light gray chart is centered barely a quarter of the way up the scale; only one guy in six is "above average" in an absolute sense. Sex appeal isn't something commonly quantified like this, so let me put it in a more familiar context: translate this plot to IQ, and you have a world where the women think 58 percent of men are brain damaged.

Now, the men on OkCupid aren't actually ugly—I tested that by experiment, pitting a random set of our users against a comparable random sample from a social network and got the same scores for both groups—and it turns out you get patterns like the above on every dating site I've seen: Tinder, Match.com, DateHookup—sites that together cover about half the single people in the United States. It just turns out that men and women perform a different sexual calculus. As *Harper's* put it perfectly: "Women are inclined to regret the sex they had, and men the sex they didn't." You can see exactly how it works in the data. I will add: the men above must be absolutely full of regrets.

A beta curve plots what can be thought of as the outcome of a large number of coin flips—it traces the overlapping probabilities of many independent binary events. Here the male coin is fair, coming up heads (which I'll equate with positive) just about as often as it comes up tails. But in our data we see that the female one is weighted; it turns up heads only once every fourth flip. A large number of natural processes, including the weather, can be modeled with betas, and thanks to some weather bug's obsessive archiving, I was able to compare our person-to-person ratings to historical climate patterns. The male outlook here is very close to the function that predicts cloud cover in New York City. The female psyche, by the same metric, dwells in a place slightly darker than Seattle.

We'll follow this thread through the first of *Dataclysm*'s three broad subjects: the data of people connecting. Sex appeal—how it changes and what creates it—will be our point of departure. We'll see why, technically, a woman is over the hill at twenty-one and the importance of a prominent tattoo, but we'll soon move beyond connections of the flesh. We'll see what tweets can tell us about modern communication, and what friendships on Facebook can say about the stability of a marriage. Profile pictures are both a boon and a curse on the Internet: they turn almost every service (Facebook, job sites, and, of course, dating) into a beauty contest. We'll take a look at what happens when OkCupid removes them for a day and just hopes for the best. Love isn't blind, though we find evidence it should be.

Part 2 then looks at the data of division. We'll begin with a close look at that prime human divide, race—a topic we can now address at the person-to-person level for the first time. Our privileged data exposes attitudes that most people would never cop to in public, and we'll see that racial bias is not only strong but consistent—repeated almost verbatim (well, numeratim) from site to site. Racism can be an interior thing too—just one man, his prejudice, and a keyboard. We'll see what Google Search has to say about the country's most hated word—and what that word has to say about the country. We'll move on to explore the divisiveness of physical beauty with a data set thousands of times more powerful than anything previously available. Ugliness has startling social costs that we are finally able to quantify. From there, we'll see what Twitter reveals about our impulse to anger. The service allows people to stay connected up to the minute; it can drive them apart just as quickly. The collaborative rage that it enables brings a new violence to that most ancient of human gatherings: the mob. We'll see if it can provide a new understanding, as well.

By the book's third section, we will have seen the data of two people interacting, for better and for worse; here we will look at the individual alone. We'll explore how ethnic, sexual, and political identity is expressed, focusing on the words, images, and cultural markers people choose to represent themselves. Here are five of the phrases most typical of a white woman:

my blue eyes
red hair and
four wheeling
country girl
love to be outside

Haiku by Carrie Underwood, or data? You make the call! We'll explore people's public words. We'll also see how people speak and act in private, with an eye toward the places where labels and action diverge: bisexual men, for example, challenge our ideas of neat identity. Next, we'll draw on a wide range of sources—Twitter, Facebook, Reddit, even Craigslist—to see ourselves in our homes, both physically and otherwise. And we'll conclude with the natural question about a book like this: how does a person maintain his privacy in a world where these explorations are possible?

Throughout, we'll see that the Internet can be a vibrant, brutal, loving, forgiving, deceitful, sensual, angry place. And of course it is: it's made of human beings. However, bringing all this information together, I became acutely aware that not everyone's life is captured in the data. If you don't have a computer or a smartphone, then you aren't here. I can only acknowledge the problem, work around it, and wait for it to go away.

I will say in the meantime that the reach of sites like Twitter and Facebook, and even my dating data, is surprisingly thorough. If you don't use many of these services yourself, this is something you might not appreciate.

Some 87 percent of the United States is online, and that number holds across virtually all demographic boundaries. Urban to rural, rich to poor, black to Asian to white to Latino, all are connected. Internet adoption is lower (around 60 percent) among the very old and the undereducated, which is why I drew my "age line" well short of old age in these pages—at fifty—and why I don't address education at all. More than 1 out of every 3 Americans access Facebook *every day*. The site has 1.3 billion accounts worldwide. Given that roughly a quarter of the world is under age fourteen, that means that something like 25 percent of adults on Earth have a Facebook account. The dating sites in *Dataclysm* have registered some 55 million American members in the last three years—as I said above, that's one account for every two single people in the country. Twitter is an especially interesting demographic case. It's a glitzy tech success story, and the company is almost single-handedly gentrifying a large swath of San Francisco. But the service itself is fundamentally populist, both in the "openness" of its platform and in who chooses to use it. For example, there's no significant difference in use by gender. People with only a high school education level tweet as much as college graduates. Latinos use the service as much as whites, and blacks use it twice as much. And then, of course, there's Google. If 87 percent of Americans use the Internet, 87 percent of them have used Google.

These big numbers don't prove I have the complete picture of anything, but they at least suggest that such a picture is coming. And in any event the perfect should not be the enemy of the better-than-ever-before. The data set we'll work with encompasses thousands of times more people than a Gallup or Pew study; that goes without saying. What's less obvious is that it's actually much more inclusive than most academic behavioral research.

It's a known problem with existing behavioral science—though it's seldom discussed publicly—that almost all of its foundational ideas were established on small batches of college kids. When I was a student, I got paid like $25 to inhale a slightly radioactive marker gas for an hour at Mass General and then do some kind of mental task while they took pictures of my brain. It won't hurt you, they said. It's just like spending a year in lane, they said. No big deal, they said. What they didn't say—and . didn't realize then—was that as I was lying there a little hungover in some kind of CAT-scanner thing, reading words and clicking buttons with my foot, I was standing in for the typical human male. My friend did the study, too. He was a white college kid just like me. I'm willing to bet most of the subjects were. That makes us far from typical.

I understand how it happens: in person, getting a real representative data set is often more difficult than the actual experiment you'd like to perform. You're a professor or postdoc who wants to push forward, so you take what's called a "convenience sample"—and that means the students at your university. But it's a big problem, especially when you're researching belief and behavior. It even has a name. It's called WEIRD research: white, educated, industrialized, rich, and democratic. And most published social research papers are WEIRD.*

Several of these problems plague my data, too. It will be a while still before digital data can scratch "industrialized" all the way off the list. But because tech is often seen as such an "elite field"—an image that many in the industry are all too willing to encourage—I feel compelled to distinguish between the entrepreneurs and venture capitalists you see on technology's public stages, making swiping gestures and spouting buzz talk into headset mikes, people who are usually very WEIRD indeed, from the users of the services themselves, who are very much normal. They can't help but be, because use of these services—Twitter, Facebook, Google, and the like—is the norm.

As for the data's authenticity, much of it is, in a sense, fact-checked because the Internet is now such a part of everyday life. Take the data from OkCupid. You give the site your city, your gender, your age, and who you're looking for, and it helps you find someone to meet for coffee or a beer. Your profile is supposed to be you, the true version. If you

---

* An article in *Slate* noted: "WEIRD subjects, from countries that represent only about 12 percent of the world's population, differ from other populations in moral decision making, reasoning style, fairness, even things like visual perception. This is because a lot of these behaviors and perceptions are based on the environments and contexts in which we grew up."

upload a better-looking person's picture as your own, or pretend to be much younger than you really are, you will probably get more dates. But imagine meeting those dates in person: they're expecting what they saw online. If the real you isn't close, the date is basically over the instant you show up. This is one example of the broad trend: as the online and offline worlds merge, a built-in social pressure keeps many of the Internet's worst fabulist impulses in check.

The people using these services, dating sites, social sites, and news aggregators alike, are all fumbling their way through life, as people always have. Only now they do it on phones and laptops. Almost inadvertently, they've created a unique archive: databases around the world now hold years of yearning, opinion, and chaos. And because it's stored with crystalline precision it can be analyzed not only in the fullness of time, but with a scope and flexibility unimaginable just a decade ago.

I have spent several years gathering and deciphering this data, not only from OkCupid, but from almost every other major site. And yet I've never quite been able to get over a nagging doubt, which, given my Luddite sympathies, pains me all the more: writing a book about the Internet feels a lot like making a very nice drawing about the movies. Why bother? That's the question of my dark hours.

∞

There's this great documentary about Bob Dylan called *Dont Look Back* that I watched a bunch back in college; my best friend, Justin, was studying film. Somewhere in the movie, at an after-party, Bob gets into an argument with a random guy about who did or who did not throw some glass thing in the street. They're both clearly drunk. The climax of the confrontation is this exchange, and it's stuck with me now for fifteen years:

DYLAN: I know a thousand cats who look just like you and talk just like you.

GUY AT PARTY: Oh, fuck off. You're a big noise. You know?

DYLAN: I know it, man. I know I'm a big noise.

GUY AT PARTY: I know you know.

DYLAN: I'm a bigger noise than you, man.

GUY AT PARTY: I'm a small noise.

DYLAN: Right.

And then someone breaks it up so they can all talk poetry. It's that kind of night. But here's the thing: rock star or no, big noises have been the sound of mankind so far. Conquerors, tycoons, martyrs, saviors, even scoundrels (especially scoundrels!)—their lives are how we've told our larger story, how we've marked our progression from the banks of a couple of silty rivers to wherever we are now. From Pharaoh Narmer in BCE 3100, the first living man whose name we still know, to Steve Jobs and Nelson Mandela—the heroic framework is how people order the world. Narmer was first on an ancient list of kings. The scribes have changed, but that list has continued on. I mean, the 1960s, power to the people and so on, is the perfect example: that's the era of Lennon and McCartney, Dylan, Hendrix, not "Guy at Party." Above all, Everyman's existence hasn't been worth recording, apart from where it intersects with a legend's.

But this asymmetry is ending; the small noise, the crackle and hiss of the rest of us, is finally making it to tape. As the Internet has democratized journalism, photography, pornography, charity, comedy, and so many other courses of personal endeavor, it will, I hope, eventually democratize our fundamental narrative. The sound is inchoate now, unrefined. But I'm writing this book to bring out what faint patterns I, and others, detect. This is the echo of the approaching train in ears pressed to the rail. Data science is far from perfect—there's selection bias and many other shortcomings to understand, acknowledge, and work around. But the distance between what could be and what is grows shorter every day, and that final convergence is the day I'm writing to.

I know there are a lot of people making big claims about data, and I'm not here to say it will change the course of history—certainly not like internal combustion did, or steel—but it will, I believe, change what history *is*. With data, history can become deeper. It can become more. Unlike clay tablets, unlike papyrus, unlike paper, newsprint, celluloid, or photo stock, disk space is cheap and nearly inexhaustible. On a hard drive, there's room for more than just the heroes. Not being a hero myself, in fact, being someone who would most of all just like to spend time with his friends and family and live life in small ways, this means something to me.

Now, as much as I'd like me and you and WhoBeefed81 to be right there on the page with the president when future works treat this decade, I imagine everyday people will always be more or less nameless, as indeed they are even here. The best data can't change that. But we all will be counted. When in ten years, twenty, a hundred, someone takes the temperature of these times and wants to understand changes—wants to see how legalizing gay marriage both drove and reflected broader acceptance of homosexuality or how village society in Asia was uprooted, then created again, within its large urban centers—inside that story, even comprising its very bones, will be data from Facebook, Twitter, Reddit, and the like. And if not, our putative writer will have failed.

I've tried to capture all this with my mash-up title. *Kataklysmos* is Greek for the Old Testament Flood; that's how the word "cataclysm" came to English. The allusion has dual resonance: there is, of course, the data as unprecedented deluge. What's being collected today is so deep it verges on bottomless; it's easily forty days and forty nights of downpour to that old handful of rain. But there's also the hope of a world transformed—of both yesterday's stunted understanding and today's limited vision gone with the flood.

This book is a series of vignettes, tiny windows looking in on our lives—what brings us together, what pulls us apart, what makes us who we are. As the data keeps coming, the windows will get bigger, but there's plenty to see right now, and the first glimpse is always the most thrilling. So to the sills, I'll boost you up.

PART 1

What Brings Us Together

P. 37

P. 63

Wooderson's Law

Up where the world is steep, like in the Andes, people use funicular railroads to get where they need to go—a pair of cable cars connected by a pulley far up the hill. The weight of the one car going down pulls the other up; the two vessels travel in counterbalance. I've learned that that's what being a parent is like. If the years bring me low, they raise my daughter, and, please, so be it. I surrender gladly to the passage, of course, especially as each new moment gone by is another I've lived with her, but that doesn't mean I don't miss the days when my hair was actually all brown and my skin free of weird spots. My girl is two and I can tell you that nothing makes the arc of time more clear than the creases in the back of your hand as it teaches plump little fingers to count: one, two, tee.

But some guy having a baby and getting wrinkles is not news. You can start with whatever the Oil of Olay marketing department is running up the pole this week and work your way back to myths of Hera's jealous rage. People have been obsessed with getting older, and with getting uglier because of it, for as long as there've been people and obsession and ugliness. "Death and taxes" are our two eternals, right? And depending on the next government shutdown, the latter is looking less and less reliable. So there you go.

When I was a teenager—and it shocks me to realize I was closer then to my daughter's age than to my current thirty-eight—I was really into punk rock, especially pop-punk. The bands were basically snottier and less proficient versions of Green Day. When I go back and listen to them now, the whole phenomenon seems supernatural to me: grown men brought together in trios and quartets by some unseen force to whine about girlfriends and what other people are eating. But at the time I thought these bands were the shit. And because they were too cool to have posters, I had to settle for arranging their album covers and flyers on my bedroom wall. My parents have long since moved—twice, in fact. I'm pretty sure my old bedroom is now someone else's attic, and I have no idea where any of the paraphernalia I collected is. Or really what most of it even looked like. I can just remember it and smile, and wince.

Today an eighteen-year-old tacks a picture on his wall, and that wall will never come down. Not only will his thirty-eight-year-old self be able to go back, pick through the detritus, and ask, "What was I thinking?," but so can the rest of us, and so can researchers. Moreover, they can do it for all people, not just one guy. And, more still, they can connect that eighteenth year to what came before and what's still to come, because the wall, covered in totems, follows him from that bedroom in his parents' house to his dorm room to his first apartment to his girlfriend's place to his honeymoon, and, yes, to his daughter's nursery. Where he will proceed to paper it over in a billion updates of her eating mush.

A new parent is perhaps most sensitive to the milestones of getting older. It's almost all you talk about with other people, and you get actual metrics at the doctor's every few months. But the milestones keep coming long after babycenter.com and the pediatrician quit with the reminders. It's just that we stop keeping track. Computers, however, have nothing better to do; keeping track is their only job. They don't lose the scrapbook, or travel, or get drunk, or grow senile, or even blink. They just sit there and remember. The myriad phases of our lives, once gone but to memory and the occasional shoebox, are becoming permanent, and as daunting as that may be to everyone with a drunk selfie on Instagram, the opportunity for understanding, if handled carefully, is self-evident.

What I've just described, the wall and the long accumulation of a life, is what sociologists call *longitudinal data*—data from following the same people, over time—and I was speculating about the research of the future. We don't have these capabilities quite yet because the Internet, as a pervasive human record, is still too young. As hard as it is to believe, even Facebook, touchstone and warhorse that it is, has only been big for about six years. It's not even in middle school! Information this deep is still something we're building toward, literally, one day at a time. In ten or twenty years, we'll be able to answer questions like . . . well, for one, how much does it mess up a person to have every moment of her life, since infancy, posted for everyone else to see? But we'll also know so much more about how friends grow apart or how new ideas percolate through the mainstream. I can see the long-term potential in the rows and columns of my databases, and we can all see it in, for example, the promise of Facebook's Timeline: for the passage of time, data creates a new kind of fullness, if not exactly a new science.

Even now, in certain situations, we can find an excellent proxy, a sort of flash-forward to the possibilities. We can take groups of people at different points in their lives, compare them, and get a rough draft of life's arc. This approach won't work with music tastes, for example, because music itself also evolves through time, so the analysis has no control. But there are fixed universals that can support it, and, in the data I have, the nexus of beauty, sex, and age is one of them. Here the possibility already exists to mark milestones, as well as lay bare vanities and vulnerabilities that were perhaps till now just shades of truth. So doing, we will approach a topic that has consumed authors, painters, philosophers, and poets since those vocations existed, perhaps with less art (though there is an art to it), but with a new and glinting precision. As usual, the good stuff lies in the distance between thought and action, and I'll show you how we find it.

I'll start with the opinions of women—all the trends below are true across my sexual data sets, but for specificity's sake, I'll use numbers from OkCupid. This table lists, for a woman, the age of men she finds *most attractive*. If I've arranged it unusually, you'll see in a second why.

a woman's age *vs. the age of the men who look best to her*

a woman's age	the age of the men who look best to her
20	23
21	23
22	24
23	25
24	25
25	26
26	27
27	28
28	29
29	29
30	30
31	31
32	31
33	32
34	32
35	34
36	35
37	36
38	37
39	38
40	38
41	38
42	39
43	39
44	39
45	40
46	38
47	39
48	40
49	45
50	46

Reading from the top, we see that twenty- and twenty-one-year-old women prefer twenty-three-year-old guys; twenty-two-year-old women like men who are twenty-four, and so on down through the years to women at fifty, who we see rate forty-six-year-olds the highest. This isn't survey data, this is data built from tens of millions of preferences expressed in the act of finding a date, and even from just following along the first few entries, the gist of the table is clear: a woman wants a guy to be roughly as old as she is. Pick an age in black under forty, and the number in gray is always very close. The broad trend comes through better when I let lateral space reflect the progression of the values in gray:

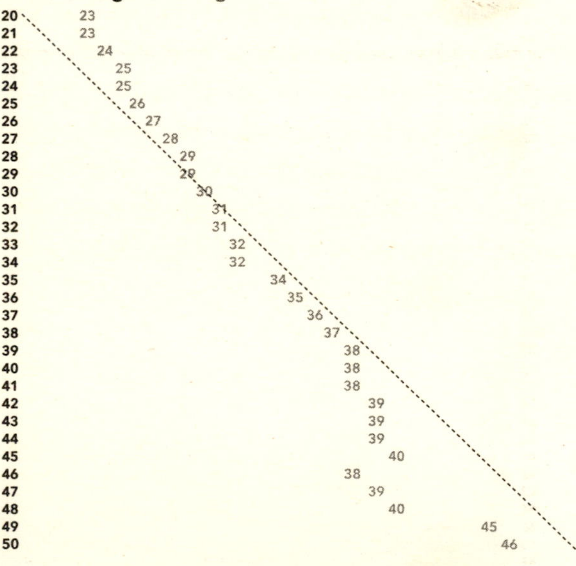

That dotted diagonal is the "age parity" line, where the male and female years would be equal. It's not a canonical math thing, just something I overlaid as a guide for your eye. Often there is an intrinsic geometry to a situation—it was the first science for a reason—and we'll take advantage wherever possible.* This particular line brings out two transitions, which coincide with big birthdays. The first pivot point is at thirty, where the trend of the **gray** numbers—the ages of the men—crosses below the line, never to cross back. That's the data's way of saying that until thirty, a woman prefers slightly older guys; afterward, she likes them slightly younger. Then at forty, the progression breaks free of the diagonal, going practically straight down for nine years. That is to say, a woman's tastes appear to hit a wall. Or a man's looks fall off a cliff, however you want to think about it. If we want to pick the point where a man's sexual appeal has reached its limit, it's there: forty.

The two perspectives (of the woman doing the rating and of the man being rated) are two halves of a whole. As a woman gets older, her stan-

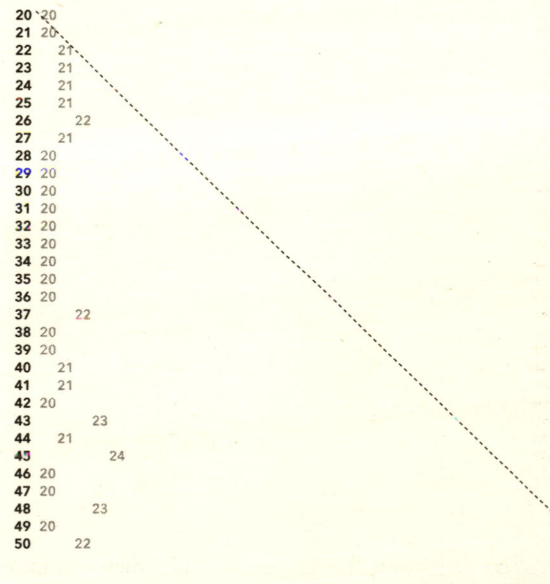

* This, in my opinion, is what distinguishes a true data visualization from, say, a plain graph or an impressionistic work of art that happens to include numbers. In a visualization, the physical space itself communicates relationships.

dards evolve, and from the man's side, the rough 1:1 movement of the gray numbers versus the black implies that as he matures, the expectations of his female peers mature as well—practically year-for-year. He gets older, and their viewpoint accommodates him. The wrinkles, the nose hair, the renewed commitment to cargo shorts—these are all somehow satisfactory, or at least offset by other virtues. Compare this to the free fall of scores going the other way, from men to women.

This graph—and it's practically not even a graph, just a table with a couple columns—makes a statement as stark as its own negative space. A woman's at her best when she's in her very early twenties. Period. And really my plot doesn't show that strongly enough. The four highest-rated female ages are twenty, twenty-one, twenty-two, and twenty-three for *every* group of guys but one. You can see the general pattern below, where I've overlaid shading for the top two quartiles (that is, top half) of ratings. I've also added some female ages as numbers in black on the bottom horizontal to help you navigate:

a man's age *vs. the age of the women who look best to him*

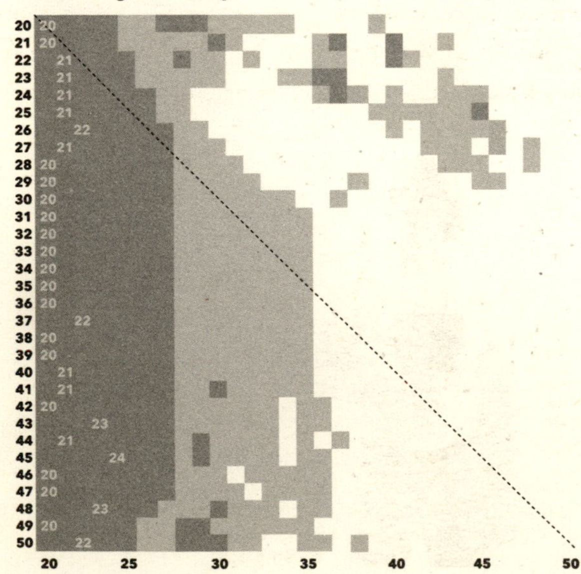

Again, the geometry speaks: the male pattern runs much deeper than just a preference for twenty-year-olds. And after he hits thirty, the latter half of our age range (that is, women over thirty-five) might as well not exist. Younger is better, and youngest is best of all, and if "over the hill" means the beginning of a person's decline, a straight woman is over the hill as soon as she's old enough to drink.

Of course, another way to put this focus on youth is that males' expectations never grow up. A fifty-year-old man's idea of what's hot is roughly the same as a college kid's, at least with age as the variable under consideration—if anything, men in their twenties are *more* willing to date older women. That pocket of middling ratings in the upper right of the plot, that's your "cougar" bait, basically. Hikers just out enjoying a nice day, then bam.

In a mathematical sense, a man's age and his sexual aims are independent variables: the former changes while the latter never does. I call this Wooderson's law, in honor of its most famous proponent, Matthew McConaughey's character from *Dazed and Confused*.

Unlike Wooderson himself, what men *claim* they want is quite different from the private voting data we've just seen. The ratings above were submitted without any specific prompt beyond "Judge this person." But when you ask men outright to select the ages of women they're looking

That's what I like about these high school girls. I get older, they stay the same age.

for, you get much different results. The gray area below is what men *tell* us they want when asked:

***a man's age** vs. the age of the women who look best to him*

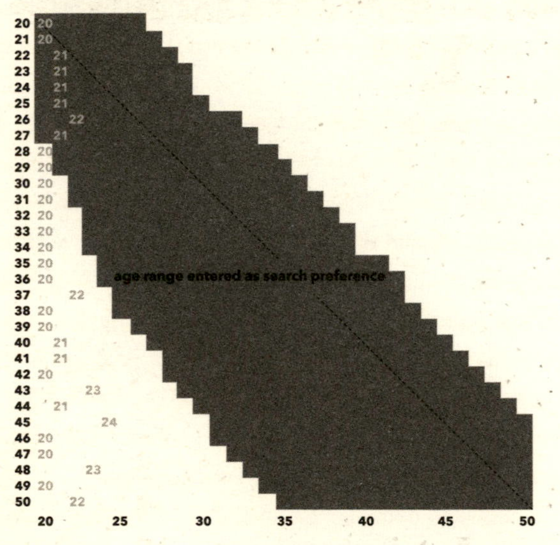

Since I don't think that anyone is intentionally misleading us when they give OkCupid their preferences—there's little incentive to do that, since all you get then is a site that gives you what you know you don't want—I see this as a statement of what men imagine they're supposed to desire, versus what they actually do. The gap between the two ideas just grows over the years, although the tension seems to resolve in a kind of pathetic compromise when it's time to stop voting and act, as you'll see.

The next plot (the final one of this type we'll look at) identifies the age with the greatest density of *contact attempts*. These most-messaged ages are described by the darkest gray squares drifting along the left-hand edge of the larger swath. Those three dark verticals in the graph's lower half show the jumps in a man's self-concept as he approaches middle age. You can almost see the gears turning. At forty-four, he's comfortable approaching a woman as young as thirty-five. Then, one year later . . . he thinks better of it. While a nine-year age difference is fine, ten years is apparently too much.

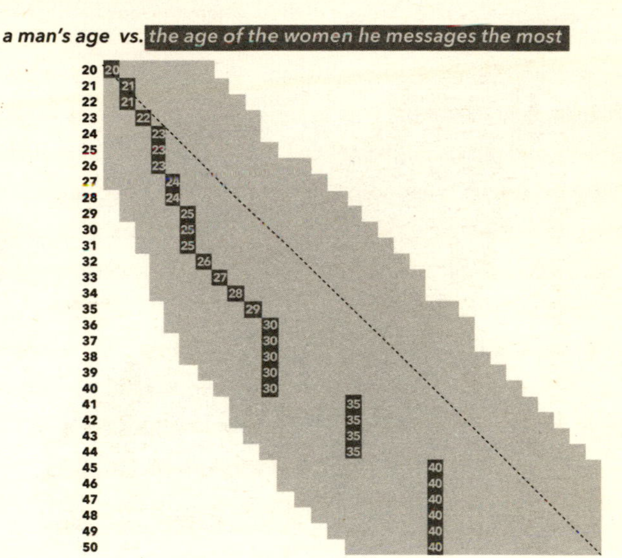

It's this kind of calculated no-man's-land—the balance between what you want, what you say, and what you do—that real romance has to occupy: no matter how people might vote in private or what they prefer in the abstract, there aren't many fifty-year-old men successfully pursuing twenty-year-old women. For one thing, social conventions work against it. For another, dating requires reciprocity. What one person wants is only half of the equation.

When it comes to women seizing the initiative and reaching out to men, because of the female-to-male attraction ratio we saw at the beginning of the chapter (1 year:1 year), plus the nonphysical motivations that push women toward older men—economics, for example—women send more, rather than fewer, messages to a man as he gets older, up until the early thirties. From there, the amount of contact declines, but no faster than the general number of available females itself is shrinking. Think about it like this: imagine you could take a typical twenty-year-old guy, who's just starting to date as an adult (definition: no SOLO cups present during at least one of courtship/consummation/breakup), and you could somehow note all the women who would be interested in him. If you could then track the whole lot over time, the main way he'll lose options from that set is when some of them just stop being single because they've paired off with someone else. In fact, his total "interested" pool would actually gain women, because as he gets older, and presumably richer and more successful, those qualities draw younger women in. In any event, his age, of itself, doesn't hurt him. Over the first two decades of his dating life, as he and the women in his pool mature, the ones who are still available will find him as desirable an option as they did when they were all twenty.

If you could do the same thing for a typical woman at twenty, you'd get a different story. Over the years, she, too, would lose men from her pool to things like marriage, but she would also lose options to time itself—as the years passed, fewer and fewer of the remaining single men would find her attractive. Her dating pool is like a can with two holes—it drains on the double.

The number of single men shrinks rapidly by age: per the US Census there are 10 million single men ages twenty to twenty-four, but only 5 million at thirty to thirty-four, and just 3.5 million at forty to forty-four. When you overlay the preference patterns we see above to those shrinking demographics, you can get a sense of how a woman's real options change over time. For a woman at twenty, this is the actual shape of the dating pool:

for a 20-year-old woman: number of men interested, by their age (20-50)

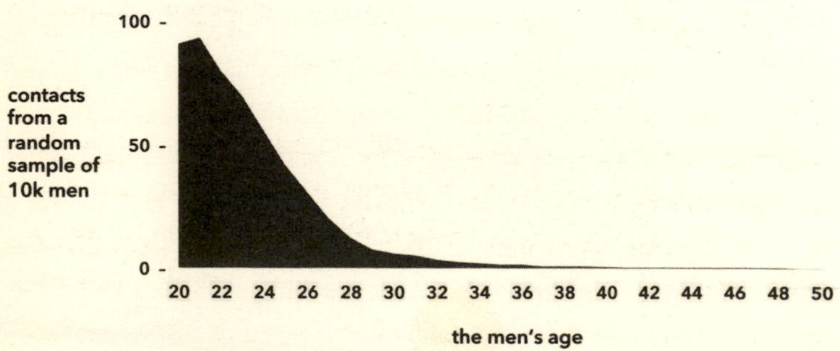

Her peers (guys in their early twenties) form the biggest component, and the numbers slope off rapidly—thirty-year-old men, for example, make up only a small part. They are less likely to actually contact someone so young, despite their privately expressed interest, and in addition many men have already partnered off by that age. By the time the woman is fifty, this is who's left (and still interested), presented on the same scale. It's Bridget Jones in charts.

for a 50-year-old woman: number of men interested, by their age (20–50)

100 -

contacts from a random sample of 10k men

50 -

0 -

20 22 24 26 28 30 32 34 36 38 40 42 44 46 48 50

the men's age

Comparing the areas, for every 100 men interested in that twenty-year-old, there are only 9 looking for someone thirty years older. Here's the full progression of charts like the two preceding, rendered from a woman's perspective for each of the ages twenty to fifty:

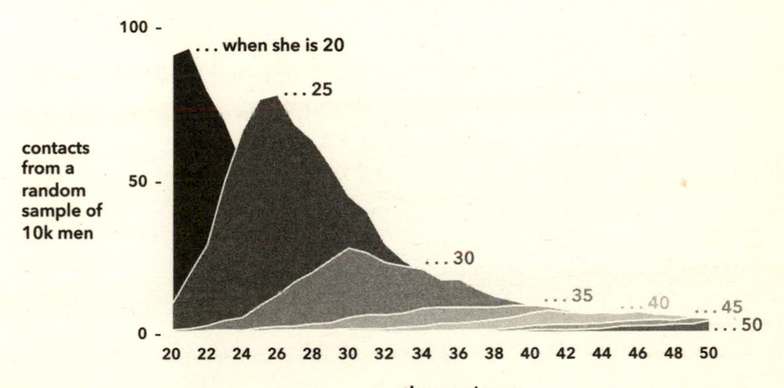

So often in my line of work, I'll see two individuals, both alone but for whatever reason not connecting. In this case, for this facet of the experience, it's two whole groups of people searching for each other at cross-purposes. Women want men to age with them. And men always head toward youth. A thirty-two-year-old woman will sign up, set her age-preference filters at 28–35, and begin to browse. That thirty-five-year-old man will come along, set his filters to 24–40, and yet rarely contact anyone over twenty-nine. Neither finds what they are looking for. You could say they're like two ships passing in the night, but that's not quite right. The men do seem at sea, pulled to some receding horizon. But in my mind I see the women still on solid ground, ashore, just watching them disappear.

2.

Death by a Thousand Mehs

In 2002, the Oscars hired the director Errol Morris to shoot a short film about why we love the movies. The Academy wanted to kick off the telecast with a rapid-fire montage of people, both celebrities and not, talking about their favorite films. My friend Justin was Morris's casting director, so he got me on the list. There was no guarantee that I'd end up in the final cut of the short, but I could do the interview on-camera and see how it went.

Having an in, I got scheduled the same day as the biggest names: Donald Trump, Walter Cronkite, Iggy Pop, Al Sharpton, Mikhail Gorbachev. Trump and Gorbachev were back to back, and somewhere out there there's a picture of the two of them, with me in the middle, photobombing before photobombing was a thing. I say "somewhere" because right after the flash, Trump snapped his fingers, and his bodyguard took Justin's camera. For his favorite movie, Trump picked *King Kong*, because he of course likes apes who try to "conquer New York." Gorbachev, through a translator whose mustache must've weighed ten pounds, chose *Gladiator.* At 2:01 in Morris's film, the wide eyes and the voice saying "*The Omen*" are mine.

Now, I like a good Antichrist movie more than most people, but I chose *The Omen* more or less at random. There are so many good movies, I'm actually not sure what my favorite one is. But I know my least favorite film with absolute certainty. *Pecker,* by John Waters. I walked out of it. Twice. I went once with some friends, couldn't deal with the mondo-trasho vibe, not to mention the exaggerated accents, and just had to leave. The next weekend, some *other* friends were going and since John Waters is a respected auteur, and hey I'm a cool guy who gets it, I figured there was at least some chance I was wrong the first time. Also I had nothing else to do. So I went again.

Such is the temporary madness of being twenty-two. I'm not saying John Waters makes objectively bad movies—they're just not for me. Or for a lot of people. And he embraces that fact, the rejection—it's practically his calling card as a director. Let me put it this way: nobody leaves *Pecker* thinking it was "meh"; either you loved it, or got the hell out after twenty minutes like I did, twice. That's by design.*

Waters's fans seem to love him all the more for being fewer in number. On OkCupid, a search through users' profile text returns more results for his name than George Lucas's and Steven Spielberg's combined. On Reddit, he has his own devoted page: */r/JohnWaters,*† and while it's not the most-trafficked URL ever, people actually put stuff there: news, old clips, questions about him, comments, and so on. There's a */r/GeorgeLucas,* too: it has one post, ever. If you enter */r/StevenSpielberg* into your address bar, you get "there doesn't seem to be anything here" from Reddit's server because, as good as his work is, no one's been enthusiastic enough to make a page. Even highly Internet-friendly directors like J. J. Abrams don't have their own page. It takes a certain special motivation to, say, make a fan site, and that motivation is often intensified by feeling like you're part of a special, embattled elect. Devotion is like vapor in a piston—pressure helps it catch.

Like many artists before and since, Waters understands exactly how it works: repelling some people draws others all the closer, and I bring him up not only because of my lifelong personal struggle with *Pecker,* but because Waters also gets the universality of the principle: it's not just true for art. He's got a lot of great quotes, but here's one that speaks right to me: "Beauty is looks you can never forget. A face should jolt, not soothe." He's completely correct, for as with music, as with movies, and as with a wide variety of human phenomena: a flaw is a powerful thing. Even at the person-to-person level, to be universally liked is to be relatively ignored. To be disliked by some is to be loved all the more by others. And, specifically, a woman's overall sex appeal is enhanced when some men find her ugly.

You can see this in the profile ratings on OkCupid. Because the site's rating system is 5 stars, the votes have more depth than just a *yes* or a

* Waters on film: "To me, bad taste is what entertainment is all about. If someone vomits while watching one of my films, it's like getting a standing ovation."

† These pages on Reddit are called subreddits. I'll explain the site and its nuances in more detail later.

no. People give degrees of opinion, and that gives us room to explore. To show this finding, we'll have to go on a short mathematical journey. These kinds of exercises are what make data science work. To put together puzzles, you have to lay out all the pieces and then just start trying things. In the absence of careful sifting, reduction, and parsimony, very little just "jumps out at you" from terabytes of raw data.

Consider a group of women with approximately the same attractiveness, let's just say the ones rated in the middle:

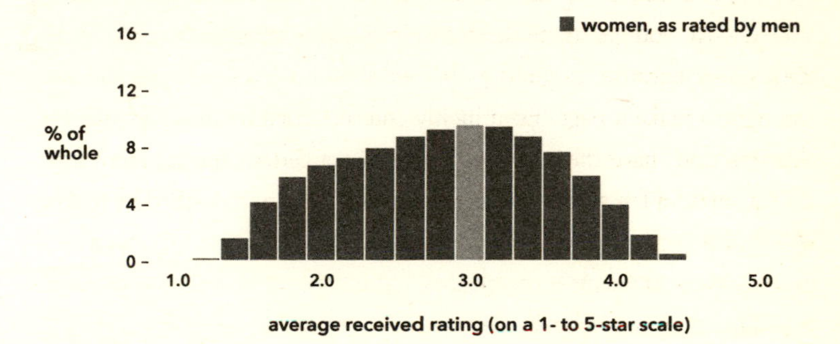

Now imagine a woman in that group and think of the many different votes men could've given her—basically think about how she ended up in the middle. There are thousands of possibilities; here are just a few I made up, combinations of 1s, 2s, 3s, 4s, and 5s, which all come to an average of 3:

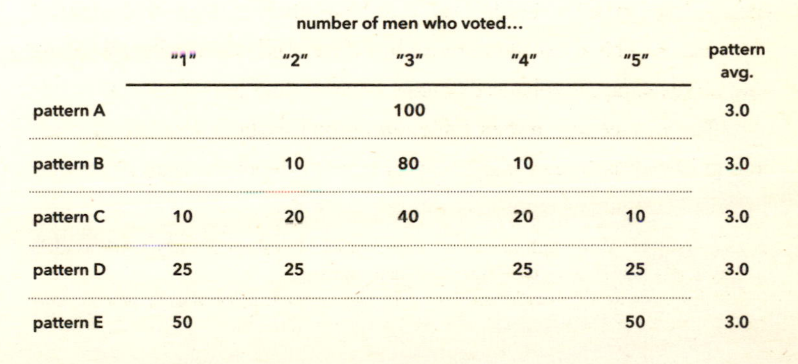

number of men who voted...

	"1"	"2"	"3"	"4"	"5"	pattern avg.
pattern A			100			3.0
pattern B		10	80	10		3.0
pattern C	10	20	40	20	10	3.0
pattern D	25	25		25	25	3.0
pattern E	50				50	3.0

As you might've noticed, the vote patterns I've chosen get more polarized as they go from Pattern A to Pattern E. Each row still averages out to that same central "3," but they express that average in different ways. Pattern A is the embodiment of consensus. There, the men who cast the votes have spoken in perfect unison: *this woman is exactly in the middle*. But by the time we get to the bottom of the table, the overall average is still centered, yet no single individual actually holds that central opinion. Pattern E shows the most extreme possible path to a middling average: for every man awarding our theoretical woman a "1," someone else gives her a "5," and the total result comes out to a "3" almost in spite of itself. That's the John Waters way.

These patterns exemplify a mathematical concept called *variance*. It's a measure of how widely data is scattered around a central value. Variance goes up the further the data points fall from the average; in the table above, it is highest in Pattern E. One of the most common applications of variance is to weigh volatility (and therefore risk) in financial markets. Consider these two companies:

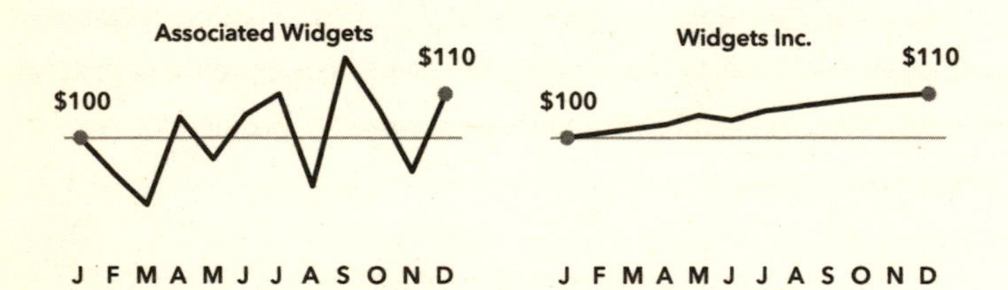

Both returned 10 percent for the year, but they are very different investments. Associated Widgets experienced large swings in value throughout the year, while Widgets Inc. grew little by little, showing consistent gains each month. Computing the variance allows analysts to capture this distinction in one simple number, and all other things being equal, investors much prefer the low score of that pattern on the right. Same return, fewer heart palpitations. Of course, when it comes to romance, heart palpitations *are* the return, and that gets to the crux of it. It turns out that variance has almost as much to do with the sexual attention a woman gets as her overall attractiveness.

In any group of women who are all equally good-looking, the number of messages they get is highly correlated to the variance: from the pageant queens to the most homely women to the people right in between, the individuals who get the most affection will be the polarizing ones. And the effect isn't small—being highly polarizing will in fact get you about 70 percent more messages. That means variance allows you to effectively jump several "leagues" up in the dating pecking order—for example, a very low-rated woman (20th percentile) with high variance in her votes gets hit on about as much as a typical woman in the 70th percentile.

Part of that is because variance means, by definition, that more people like you *a lot* (as well as dislike you a lot). And those enthusiastic guys—let's just call them the fanboys—are the ones who do most of the messaging. So by pushing people toward the high end (the 5s), you get more action.

But the negative votes themselves are part of the story, too. They drive some of the attention on their own. For example, the real patterns exemplified by C and D below get about 10 percent more messages than the ones shown in A and B, even though the top two women are rated far better overall:

	number of men who voted...					
	"1"	**"2"**	**"3"**	**"4"**	**"5"**	**pattern avg.**
woman A	**2**	**22**	**27**	**29**	**20**	**3.4**
woman B	**10**	**13**	**31**	**28**	**18**	**3.3**
woman C	**32**	**22**	**12**	**16**	**18**	**2.7**
woman D	**47**	**13**	**6**	**19**	**15**	**2.4**

I've been talking about messages as if they're an end unto themselves, but on a dating site, messages are the precursor to outcomes like in-depth conversations, the exchange of contact information, and eventually in-person meetings. People with higher variance get more of all these things, too. So, for example, woman D would have about 10 percent more conversations, 10 percent more dates, and, likely, 10 percent more sex than woman A, even though in terms of her absolute rating she's much less attractive.

Moreover, the men giving out those 1s and 2s are not themselves hitting on the women—people practically never contact someone they've rated poorly.* It's that having haters somehow induces everyone else to want you more. People *not* liking you somehow brings you more attention entirely on its own. And, yes, in his underground castle, Karl Rove smiles knowingly, petting an enormous toad.

It only adds to the mystery of the phenomenon that OkCupid doesn't publish raw attractiveness scores (or a variance number, of course) for anyone on the site. Nobody is consciously making decisions based on this data. But people have a way of feeling the math behind things, whether they're aware of it or not, and here's what I think is going on. Suppose a guy is attracted to a woman he knows is unconventional-looking. Her very unconventionality implies that some other men are likely turned off; it means less competition. Having fewer rivals increases his chances of success. I can imagine our man browsing her profile, circling his cursor, thinking to himself: *I bet she doesn't meet many guys who think she's awesome. In fact, I'm actually into her for her quirks, not in spite of them. This is my diamond in the rough*, and so on. To some degree, her very unpopularity is what makes her attractive to him. And if our browsing guy was at all on the fence about whether to actually introduce himself, this might make the difference.

Looking at the phenomenon from the opposite angle—the low-variance side—a relatively attractive woman with consistent scores is some-

* Only 0.2 percent of the messages on the site are sent by users to a person to whom they awarded fewer than 3 stars.

one any guy would consider conventionally pretty. And she therefore might seem to be more popular than she really is. Broad appeal gives the impression that other guys are after her, too, and that makes her incrementally less appealing. Our interested but on-the-fence guy moves on.

This is my theory at least. But the idea that variance is a positive thing is fairly well established in other arenas. Social psychologists call it the "pratfall effect"—as long as you're generally competent, making a small, occasional mistake makes people think you're *more* competent. Flaws call out the good stuff all the more. This need for imperfection might just be how our brains are put together. Our sense of smell, which is the most connected to the brain's emotional center, prefers discord to unison. Scientists have shown this in labs, by mixing foul odors with pleasant ones, but nature, in the wisdom of evolutionary time, realized it long before. The pleasant scent given off by many flowers, like orange blossoms and jasmine, contains a significant fraction (about 3 percent) of a protein called indole. It's common in the large intestine, and on its own, it smells accordingly. But the flowers don't smell as good without it. A little bit of shit brings the bees. Indole is also an ingredient in synthetic human perfumes.

You can see a public implementation, as it were, of the OkCupid data in the rarefied world of modeling. The women are all professionally gorgeous—5 stars out of 5, of course. But even at that high level it's still about distinguishing yourself through imperfection. Cindy Crawford's career took off after she stopped covering her mole. Linda Evangelista had the severe hair—you can't say it made her *prettier*, but it did make her far more interesting. Kate Upton, at least according to the industry standard, has a few extra pounds. Pulling a few examples from the data set, perhaps ones that are more relatable than swimsuit models, will help you see how it works for a normal person. On the following page are six women, all with middle-of-the-road overall scores, but who tend to get extreme reactions either way: lots of Yes, lots of No, but very little Meh:

Thanks to each of them for having the confidence to agree to be displayed and discussed here. What you see in the array is what you get throughout the corpus. These are people who've purposefully abandoned the middle road: with body art, a snarky expression, or by eating a grilled cheese like a badass. And you find many relatively normal women with an unusual trait: like the center woman in the bottom row, whose blue hair you can't see in black and white. And you especially see women who've chosen to play up their particular asset/liability. If you can pull off, say, a 3.3 rating despite the extra pounds or the people who hate tattoos or whatever, then, literally, more power to you.

So at the end of it, given that everyone on Earth has some kind of flaw, the real moral here is: be yourself and be brave about it. Certainly trying to fit in, just for its own sake, is counterproductive. I know this is dangerously close to the kind of thing that gets put on a quilt, and quilts, being the PowerPoint presentations of an earlier time, are the opposite of science. It also sounds a lot like the advice a mother gives, along with a pat on the head, to her big-nosed and brace-faced son when he's fourteen and can't figure out why he isn't more popular. But either way, there it is, in the numbers. Like I said, people can feel the math behind things, especially, thankfully, moms. I just wish she'd told me that by ninth grade bears aren't cool.

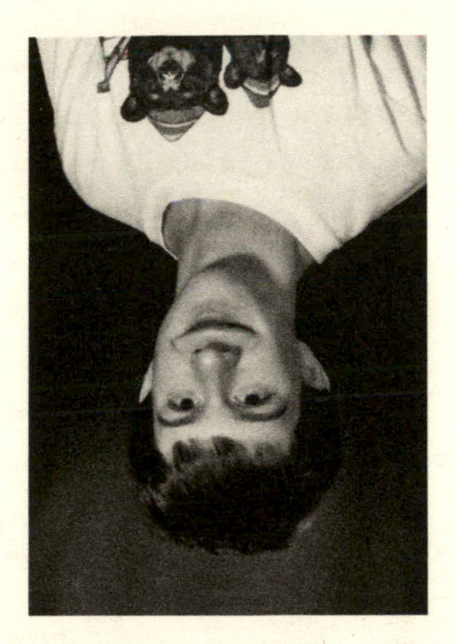

Writing on the Wall

Nostalgia used to be called *mal du Suisse*—the Swiss sickness. Their mercenaries were all over Europe and were apparently notorious for wanting to go home. They would get misty and sing shepherd ballads instead of fighting, and when you're the king of France with Huguenots to burn, songs won't do. The ballads were banned. In the American Civil War nostalgia was such a problem it put some 5,000 troops out of action, and 74 men died of it—at least according to army medical records. Given the circumstances, being sad to death is actually kind of understandable, but then again, this was also the time of leeches and the bonesaw, so who knows what was really going on. It's interesting to think that in those days, many of the people who left home did so to go to war—much of the early literature on nostalgia, which was seen then as a bona fide disease, mentions soldiers. In that sepia-toned way I can't help but think about the past, I like to imagine scientists in 1863, on either side of the Potomac, working furiously against the clock to develop the ultimate war-ending superweapon: high school yearbooks.

I actually don't even know if they have high school yearbooks anymore. It's hard to see why you'd need one now that Facebook's around, although according to the company's last quarterly report, people under eighteen aren't using Facebook as much as they used to. So maybe the kids need the printed copy again, I don't know.* But however teenagers are staying in touch—whether it's through Snapchat or WhatsApp or Twitter—I'm positive they're doing it with words. Pictures are part of the appeal of all of these services, obviously, but you can only say so much without a keyboard. Even on Instagram, the comments and the captions are essential—the photo after all is just a few inches square. But the words are the words are the words. They're still how feelings come across and how connections are made.

In fact, for all the hand-wringing over technology's effect on our culture, I am certain that even the most reticent teenager in 2014 has written far more in his life than I or any of my classmates had back in the early

* Definition of true ignorance: getting your "what the kids are into" intel from the Securities and Exchange Commission.

'90s. Back then, if you needed to talk to someone you used the phone. I wrote a few stiff thank-you notes and maybe one letter a year. The typical high school student today must surpass that in a morning. The Internet has many regrettable sides to it, but that's one thing that's always stood it in good stead with me: it's a writer's world. Your life online is mediated through words. You work, you socialize, you flirt, all by typing. I honestly feel there's a certain epistolary, Austenian grandness to the whole enterprise. No matter what words we use or how we tap out the letters, we're writing to one another more than ever. Even if sometimes

```
dam gerl
```

is all we have to say.

∞

Major Sullivan Ballou was one of the soldiers in the Union army, on the Potomac, suffering, and homesick. Early in Ken Burns's *The Civil War,* a narrator reads his farewell letter to his wife, to his "very dear Sarah," and it's a moving and important moment in the film. The Major was writing from camp before the first large battle of the war, and he was mortally wounded days later. His words were the last his family would ever hear from him, and they drove home the greater sorrow the nation would face in the years to come. Because of the exposure, the Ballou letter has become one of the most famous ever written—when I search for "famous letter," Google lists it second. It's a beautiful piece of writing, but think of all the other letters that will never be read aloud, that were burned, lost in some shuffle, or carried off by the wind, or that just moldered away.

Today we don't have to rely on the lucky accident of preservation to know what someone was thinking or how he talked, and we don't need the one to stand in for the many. It's all preserved, not just one man to one wife before one battle, but all to all, before and after and even in the middle of each of our personal battles. You can find readings of the Ballou letter on YouTube, and many of the comments are along the lines of "They just don't make them like that anymore." That's true. But what they, or rather we, are making offers a richness and a beauty of a different kind: a poetry not of lyrical phrases but of understanding. We are at the cusp of momentous change in the study of human communication and what it tries to foster: community and personal connection.

When you want to learn about how people write, their unpolished, unguarded words are the best place to start, and we have reams of them. There will be more words written on Twitter in the next two years than contained in all books ever printed. It's the epitome of the new communication: short and in real time. Twitter was, in fact, the first service not only to encourage brevity and immediacy, but to require them. Its prompt is "What's happening?" and it gives users 140 characters to tell the world. And Twitter's sudden popularity, as much as its sudden redefinition of writing, seemed to confirm the fear that the Internet was "killing our culture." How could people continue to write well (and even think well) in this new confined space—what would become of a mind so restricted? The actor Ralph Fiennes spoke for many when he said, "You only have to look on Twitter to see evidence of the fact that a lot of English words that are used, say, in Shakespeare's plays or P. G. Wodehouse novels . . . are so little used that people don't even know what they mean now."

Even basic analysis shows that language on Twitter is far from a degraded form. Below, I've compared the most common words on Twitter against the Oxford English Corpus—a collection of nearly 2.5 billion words of modern writing of all kinds—journalism, novels, blogs, papers, everything. The OEC is the canonical census of the current English vocabulary. I've charted only the top 100 words out of the tens of thousands that people use, which may seem like a paltry sample, but roughly half of all writing is formed from these words alone (both on Twitter and in the OEC). The most important thing to notice on Twitter's list is this: despite the grumblings from the weathered sentinels atop Fortress English, there are only two "netspeak" entries—*rt*, for "retweet" and *u*, for "you"—in the top 100. You'd think that contractions, grammatical or otherwise, would be staples of a form that only allows a person 140 characters, but instead people seem to be writing around the limitation rather than stubbornly through it. Second, when you calculate the average word length of the Twitter list, it's *longer* than the OEC's: 4.3 characters to 3.4. And look beyond length to the content of the Twitter vocabulary. I've highlighted the words unique to it in order to make the comparison easier:

	OEC	Twitter		OEC	Twitter
1	the	to	51	when	back
	be	a		make	an
	to	i		can	see
	of	the		like	more
	and	and		time	by
	a	in		no	today
	in	you		just	twitter
	that	my		him	or
	have	for		know	as
10	I	on	60	take	make
	it	of		people	who
	for	it		into	got
	not	me		year	here
	on	this		your	want
	with	with		good	need
	he	at		some	happy
	as	just		could	too
	you	so		them	u
	do	be		see	best
20	at	rt	70	other	people
	this	out		than	some
	but	that		then	they
	his	have		now	life
	by	your		look	there
	from	all		only	think
	they	up		come	going
	we	love		its	why
	say	do		over	he
	her	what		think	really
30	she	like	80	also	way
	or	not		back	come
	an	get		after	much
	will	no		use	only
	my	good		two	off
	one	but		how	still
	all	new		our	right
	would	can		work	night
	there	if		first	home
	their	day		well	say

40	what	now	90	way	great
	so	time		even	never
	up	from		new	work
	out	go		want	would
	if	how		because	last
	about	we		any	first
	who	will		these	over
	get	one		give	take
	which	about		day	its
	go	know		most	better
50	me	when	100	us	them

While the OEC list is rather drab, lots of helpers and modifiers—workmanlike language to get you to some payoff noun or verb—on Twitter, there's no room for functionaries; every word's gotta be boss. So you see vivid stuff like:

love
happy
life
today
best
never
home

. . . make the top 100 cut. Twitter actually may be improving its users' writing, as it forces them to wring meaning from fewer letters—it embodies William Strunk's famous dictum, *Omit needless words,* at the keystroke level. A person tweeting has no option but concision, and in a backward way the character limit actually explains the slightly longer word length we see. Given finite room to work, longer words mean fewer spaces between them, which means less waste. Although the thoughts expressed on Twitter may be foreshortened, there's no evidence here that they're diminished.

Mark Liberman, a professor of linguistics at the University of Pennsylvania, concluded much the same thing: in a direct response to Mr. Fiennes, he calculated the typical word length in *Hamlet* (3.99) and in a collection of Wodehouse's stories (4.05) and found them both less than the length in his

Twitter sample (4.80).* He's just one of many comparative linguists who've begun mining Twitter's data. A team at Arizona State was able to reach beyond word count and length, and into the sentiment and style of the writing, and they found several surprising things: first, Twitter does not change how a person writes. Among the many examples they tracked, if a writer uses "u" for the second person in e-mails or text messages, she will also use it on Twitter. But, likewise, if she generally spells out "you," she does so everywhere—on Twitter, in texts, in e-mail, and so on. The decision to refer to the first-person singular as "I" or "i" follows the same pattern. That is, a person's style doesn't change from medium to medium; there is no "dumbing down." You write how you write, wherever you write. The linguists also measured Twitter's lexical density, its proportion of content-carrying words like verbs and nouns, and found it was not only higher than e-mail's, but was comparable to the writing on *Slate*, the control used for magazine-level syntax. Everything points to the same conclusion: that Twitter hasn't so much altered our writing as just gotten it to fit into a smaller place. Looking through the data, instead of a wasteland of cut stumps, we find a forest of bonsai.

This kind of in-depth analysis (lexical density, word frequency) hints at the real nature of the transformation under way. The change Twitter has wrought on language itself is nothing compared with the change it is bringing to the *study* of language. Twitter gives us a sense of words not only as the building blocks of thought but as a social connector, which indeed has been the purpose of language since humanity hunched its way across the Serengeti. And unlike older media, Twitter gives us a way to track those bonds on an individual level. You can see not only what a person says, but who she says it to, when, and how often. Comparative linguists have long traced group commonalities through language. Basic words often share common sounds (like *tres, trois, drei, three,* and *thran,* from Spanish, French, German, English, and India's Gujarati) and those stems have given us a sense of the movements of genes and culture across the face of time.

* Liberman (and I) stripped URLs and the special signs @ and # from the analysis, so these numbers aren't artificially boosted by "nonword" material.

Researchers are already grouping people by the language they use on Twitter. Here I've excerpted an early attempt to find the tribes and emerging dialects—this is from a corpus of 189,000 tweeters sending 75 million tweets among them.

subgroups on Twitter by messaging pattern

example words	characteristic speech	percent of sample
nigga, poppin, chillin	shortened endings (e.g., -er => -a or -ing => -in)	14
tweetup, metrics, innovation	tech buzzspeak	12
inspiring, webinar, affiliate, tips	marketing self-help	11
etsy, adorable, hubby	crafting lingo	5
pelosi, obamacare, beck, libs	partisan talking points	4
bieber, pleasee, youu, <33	lengthened endings (repeated last letter)	2
anipals, pawesome, furever	animal-based puns	1
kstew, robsessed, twilighters	amalgamations/puns around the *Twilight* movies	1

It's important to note that the study grouped users by their words alone, who they messaged, and what they wrote—these language clusters were not determined a priori. The top-listed group is in fact the largest the researchers detected, and it also happens to be the most voluble (sending the most tweets per capita) as well as the most insular. Some 90 percent of the tweets sent by the group are directed within it, and its users' language is most strongly "characteristic"—half of their 100 most representative words fit the "shortened endings" pattern. Throughout the list you see groups typified by slang, pop culture references, jargon, goofy puns—people drawn together by special ways of speaking, and it's exactly the kind of language (and information) that until now has been lost to history. Like knowing a man's last words to his wife, knowing how people talk among friends gives you a much deeper sense of who they are. Technocrats, political wonks, marketing gurus, the robsessed; it will be interesting in the coming years to see how all these groups merge and recombine, and we'll be able to track it all through their text.

Once language and data come together, it's that extra dimension, time, that's so compelling. Going forward, services like Twitter will be indispensable. Looking back, Google Books is working to repair our historical blind

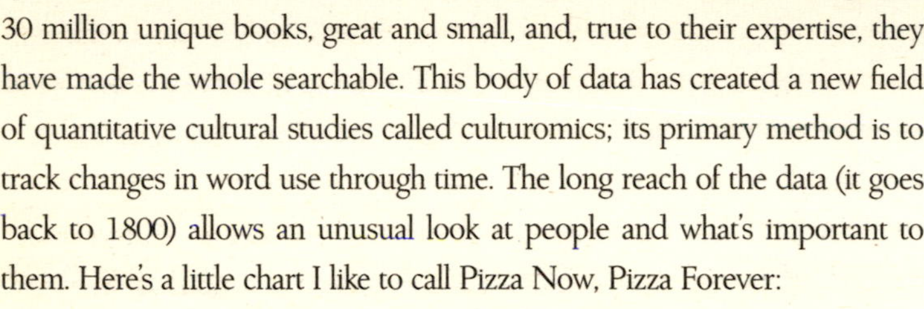

spot: in collaboration with libraries around the world, they have digitized 30 million unique books, great and small, and, true to their expertise, they have made the whole searchable. This body of data has created a new field of quantitative cultural studies called culturomics; its primary method is to track changes in word use through time. The long reach of the data (it goes back to 1800) allows an unusual look at people and what's important to them. Here's a little chart I like to call Pizza Now, Pizza Forever:

frequency of written mentions of selected foods, 1800-2008

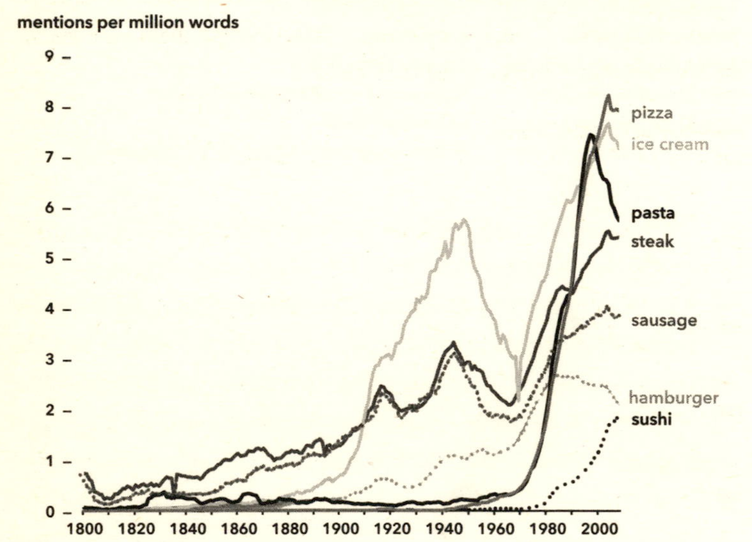

You can read bits of nonculinary history in the data, too. "Ice cream" took off in the 1910s—right when GE introduced the powered home icebox. See the nosedive "pasta" took in the late '90s? The Atkins diet became popular. During world wars, we like red meat. These are light applications of a technique that can have deep reach into our collective psyche.* Word

* The data in Google Books accounts for the fact that more books are published now than were published in, say, the nineteenth century. It samples a set number of books from each year. So though both the charts here happen to show increased mentions of their subject terms over time, that truly is a function of increased interest. Not all terms follow that pattern—"God," for example,

frequencies can even show how we perceive abstractions, like the passage of time—something very difficult to investigate directly. Asking a person what "ten years" means is like asking him or her to describe a color—you get impressionism where you're looking for facts. But looking at writing over time gives us a sense.

The data shows that with each passing year, we're getting more wrapped up in the present. For example, written mentions of the year 1850 peaked (in 1851) at roughly 35 instances for every million words written. Mentions of the year 1900 peaked at 58 per million. Mentions of recent years peak at roughly three times that. Here are the trajectories of the fifty-year benchmarks in the data set:

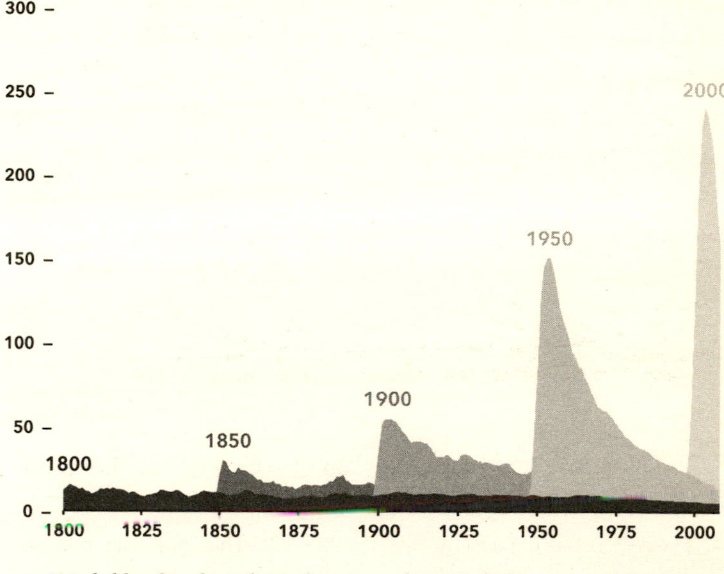

Work like this, based on the printed word, helps us understand our larger culture. Twitter lets us see groups coming together within it. But books

has been in steady decline for decades and is now used only about a third as much in American writing as it was in the early 1800s. The researchers Jean-Baptiste Michel and Erez Lieberman Aiden coined the term "culturomics" in their paper "*Quantitative Analysis of Culture Using Millions of Digitized Books*." My charts and findings here are adapted from their work.

and tweets both are one-to-many forms of communication, and, often, like Major Ballou's, our most important words are expressed one-to-one. Users on OkCupid exchange about 4 million messages a day. Of course, they do so with a special purpose—dating—but the interface provides no specific prompt and enforces no limit on what or how much anyone types. Think of it as Gmail for strangers: the communication on the site is about two people getting to know each other; the romance comes much later, offline. Outside researchers rarely get to work with private messages like this—it's the most sensitive content users generate and even anonymized and aggregated, message data is rarely allowed out of the holiest of holies in the database. But my unique position at OkCupid gives us special access.

First, the site's decade of history lets us see how technology has altered how people communicate. OkCupid has records from the pre-smartphone, pre-Twitter, pre-Instagram days—hell, it was online when Myspace was still a file storage service. Judging by messaging over all those years, the broad writing culture is indeed changing, and the change is driven by phones. Apple opened their app store in mid-2008, and OkCupid, like every major service, quickly launched an app. The effect on writing was immediate. Users began typing on keyboards smaller than their palm, and message length has dropped by over two-thirds since:

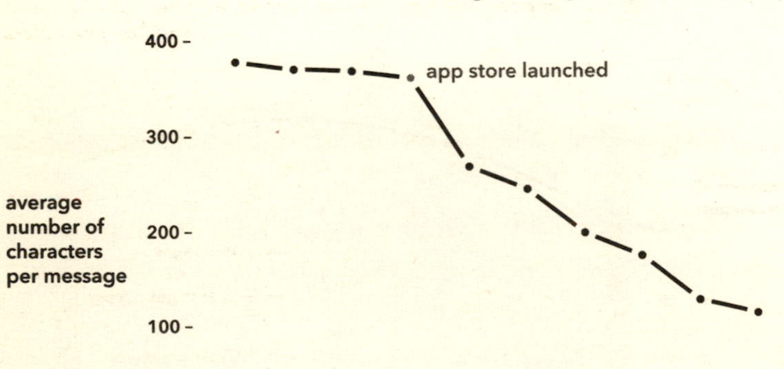

The average message is now just over 100 characters—Twitter-sized, in fact. And in terms of effect, it seems readers have adapted. The best messages, the ones that get the highest response rate, are now only 40 to 60 characters long.

message length vs. response rate earned

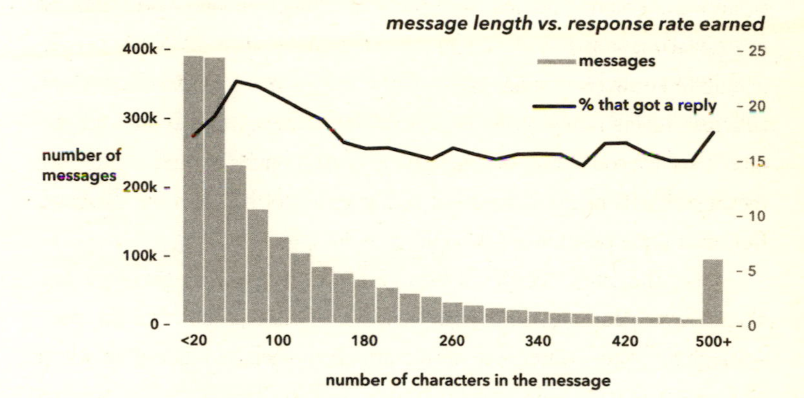

By considering only messages of a certain length, and then asking how many seconds the message took to compose, we can get a sense of how much revision and effort translates into better results. Below are messages between 150 and 300 characters, plotted against how long they took to write. As you can see, taking your time helps, up to a point. But the downward bend of the trend lines is a wingman in numbers, saying *don't overthink it!*

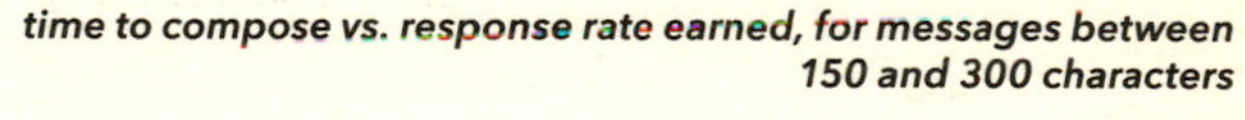

time to compose vs. response rate earned, for messages between 150 and 300 characters

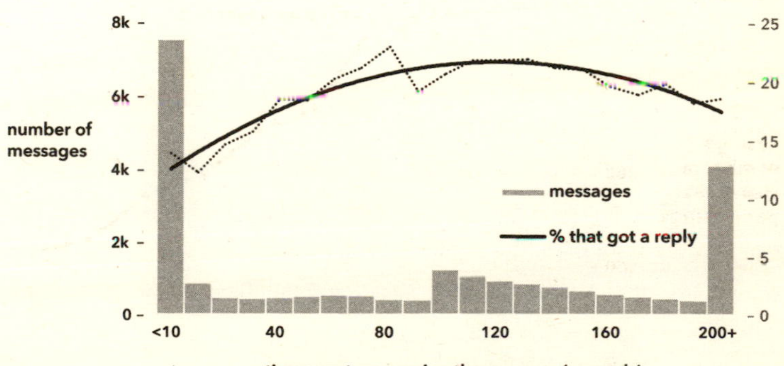

Now, the first vertical on the left, the messages that took no more than ten seconds to write, represents an inordinate amount of the whole and should raise some eyebrows. It raised mine for sure, and at this point I'm so jaded my face is frozen—Botox has nothing on ten years working at a dating site. How are so many people typing messages that long that quickly? The short answer is, they're not, and here's how I know.

Below is a scatter chart of 100,000 messages, with the number of characters *typed* plotted against characters *actually sent*.* Because there's a wide range of counts, running from 1 all the way to almost 10,000, this plot is logarithmic:

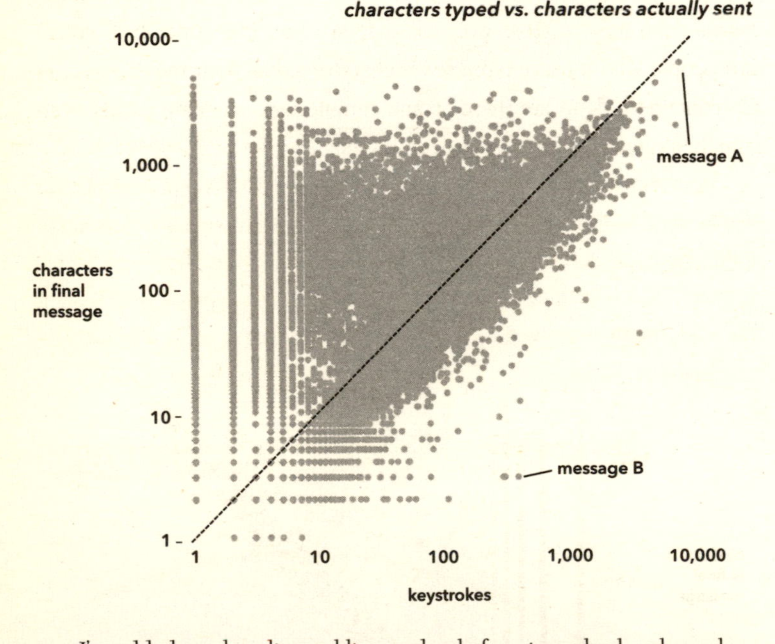

I've added another diagonal line, and as before, it marks the place where the two axes are equal—meaning that for the gray dots along it, the text matched the keystrokes that went into it. Essentially, the sender typed what was on his mind and hit Send, no backspace, no edits. Therefore we know

* I captured the characters typed through a script introduced for this chapter.

that message A, in the upper-right corner, was typed more or less in a headlong rush, with almost no revision. Going back to the logs, I found it took the sender 73 minutes and 41 seconds to hammer out those 5,979 characters of hello—his final message was about as long as four pages in this book. He did not get a reply. Neither did the gentleman sender of B, who wins the Raymond Carver award for labor-intensive brevity. He took 387 keystrokes to get to "Hey."

But these are the examples at the extremes. The broad gist of the scatter plot is: as you approach the diagonal, the messages show less revision. Move toward the bottom right, you get heavy editing, toward the upper left, you get . . . physical impossibility. Our chart's geometry means that as soon as you cross over the diagonal into the upper half, you're into people who must've *typed* fewer characters than their messages actually contained. Who are these arcane summoners, wringing words from thought alone? They are the cut and pasters, and they are legion.

We can clarify the graph by making each dot 90 percent transparent. This lets you see the real density underneath. It's like we're taking an X-ray of the data, and in so doing, we see the bones:

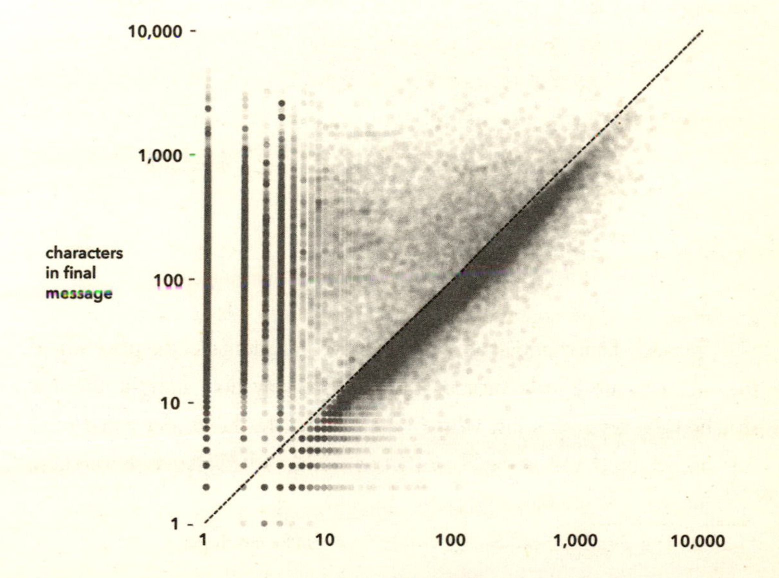

That dense band of dots running just below the diagonal is the writing-from-scratch guys. It's surprisingly compact. There is, of course, the hard upper boundary of the line, which separates the from-scratch messages from the pasted ones, like a border between warring factions. But the band's lower boundary is almost as crisp. There appears to be a natural limit to how much effort a person is willing to put into a message. If you do the arithmetic, it's 3 characters typed for every 1 in the finished product.

Above the diagonal are the people who decided that kind of effort was too much. That diffusion of dots in the upper-left center is all the people who pasted a templated message and made a few edits to it. Here the logarithmic nature of the chart can fool you—even just a small amount over that central line means most of the content in the message is stock. Running up the left side, you see the dense vertical lines, the ruts. Those are the messages that were "typed" with just a few keystrokes. There are a lot of them—all told, 20 percent of the sample registered 5 or fewer keystrokes. These writers settled on something they like or that works, and they went with it. It's not spam in the way we normally use that word—OkCupid is quick to get fake or bot accounts off the site. These are real people's attempts at contact, essentially memorized digital pickup lines. Many are about as lazy and mundane as you'd expect: "Hey you're cute" or "Wanna talk?"—just digital equivalents of "Come here often?" But some of the repeated messages are so idiosyncratic it's hard to believe they would even apply to multiple people. Here's one, presented exactly as typed:

> I'm a smoker too. I picked it up when backpacking in May. It used to be a drinking thing, but now I wake up and fuck, I want a cigarette. I sometimes wish that I worked in a Mad Men office. Have you seen the Le Corbusier exhibit at MoMA? It sounds pretty interesting. I just saw a Frank Gehry (sp?) display last week in Montreal, and how he used computer modelling to design a crazy house in Ohio.

That's the whole message—the sender was trying to pick up women who smoked and were into art. The unstudied "(sp?)" is my favorite flourish. Forty-two different women got this same message.

Sitewide, the copy-and-paste strategy underperforms from-scratch messaging by about 25 percent, but in terms of effort-in to results-out it always wins: measuring by replies received per unit effort, it's many times more efficient to just send everyone roughly the same thing than to compose a new message each time. I've told people about guys copying and pasting, and the response is usually some variation of "That's so lame." When I tell them that boilerplate is 75 percent as effective as something original, they're skeptical—surely almost everyone sees through the formula. But this last message is an example of a replicated text that's impossible to see through, and, in a fraction of the time it would've taken him otherwise, the sender got five replies from *exactly* the type of woman he was looking for. And let me tell you something. Nearly every single thing on my desk, on my person, probably in my entire home, was made in a factory alongside who knows how many copies. I just fought a crowd to pick up my lunch, which was a sandwich chosen from a wall of sandwiches. Templates work. Our social-smoking architecture-loving backpacker is just doing what people have always done: harnessing technology. In this case his innovation is using a few keyboard shortcuts to save himself some time.

As we've seen, phones and services like Twitter demand their own adaptations. The eternal here is that writing, like life itself, abides. It changes form, it replicates in odd ways, it finds unexpected niches . . . it even, like anything alive, occasionally stinks. But realize this: we are living through writing's Cambrian explosion, not its mass extinction. Language is more varied than ever before, even if some of it is directly copied from the clipboard—variety is the *preservation* of an art, not a threat to it. From the high-flown language of literary fiction to the simple, even misspelled, status update, through all this writing runs a common purpose. Whether friend to friend, stranger to stranger, lover to lover, or author to reader, we use words to connect. And as long as there is a person bored, excited, enraged, transported, in love, curious, or missing his home and afraid for his future, he'll be writing about it.

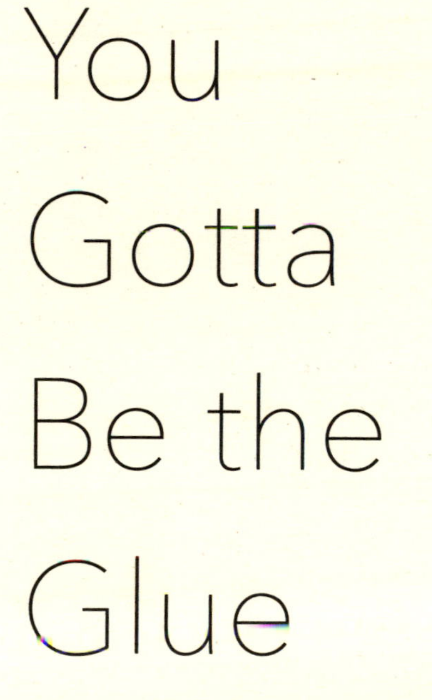

4. You Gotta Be the Glue

A major drawback to data from dating sites is that it tells you next to nothing about people actually going on dates. Once people are together in person, they don't need messages or ratings or any of that. It's an irony both in the data set and in the job itself—you do it right and the customers leave. In pairs, no less!

Where they go, of course, is into the real world, into a bar, into daylight, into the flesh. They depart the easily quantified world of bits and pixels and enter, in short, each other's lives. Think about the progression of a young relationship. Two people meet for the first time in person. Talk, drink, get to know each other. Next, if there is a next, is the apartments. The unfamiliar number on the door, a brass handle where yours is steel. The strange but pleasant smell of another person's sheets. Shampoos in the shower, used, but new to you. Loganberry: Okay, why not? Back at your place next time, she opens the fridge, and it's just . . . mustards. Sorry. We've all been there in someone's bedroom, in the den, amidst mementos of events and people we don't remember, wondering first at the tchotchkes themselves and then soon enough at how surprisingly *yours* something like the Ponderosa Invitational Swim Meet (third-place cup, 1985) can become, in spite of the fact—or is it because?—you only know it through her.

You meet the friends. The best friend. The other best friend. The *other* other best friend, like, for real, they've known each other forever. Enough drinks, the right kind of people, they become your friends, too. Acquaintances, coworkers filter into the picture, some in passing, some on purpose. Finally, maybe, if it's really turning into something, come the parents. You relate some fancier version of your life story, parts of which the two of you can tell together, because you're that familiar—step away from the table for a second, and the parents know more about you than when you left. Settling back into your chair: "M tells me that . . ." and it's the perfect setup for one of your favorite stories. Two lives are merging. And then, often, and often suddenly, it's back to the beginning with someone else.

We've had a look so far at the ways two people come together in the first blush of attraction. I'm not sure a computer will ever capture their path to full togetherness, but we do have a picture of their lives once they get there. That pattern of a couple together, the enmeshing of what's come to be called their "social graphs," is now well documented.

I have 384 friends on Facebook, and here they are. I'm the dot in the middle; my wife, Reshma, is in black at about three o'clock. Everyone's connections to everyone else are shown by the gray lines:

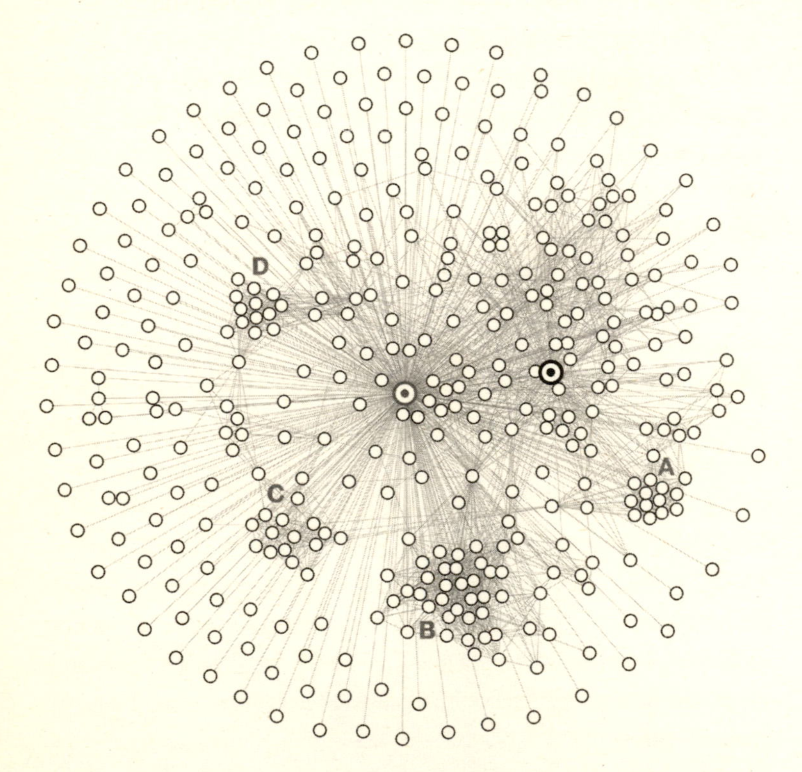

Though the groups of my friends are nicely clustered, this plot wasn't arranged by hand—my able research assistant, James Dowdell, wrote special software to create it. The dots come together based on their number of shared connections. Think of them as little bits of iron dust magnetized by the POWER OF FRIENDSHIP, and then dropped on a tabletop to settle into place. Even though I don't use Facebook for much of anything besides the highly circular task of accepting Facebook friend requests, you can see all the sides of my life in there. My very tight-knit set of in-laws, as near to overlapping as the software would allow, is A; the people I went to high school with are B; my coworkers are C; my gamer friends, D. You can even read my once and future career as a musician in the graph. I spent years touring in a band, and those singleton dots all along the left perimeter are primarily people I met on the road. Their bond to one another is our music, invisible to algorithms.

Let me expand the graph to include Reshma's connections as well, to show the scope of our network as a couple. The connections we share, our mutual friends, are the circles filled in with gray.

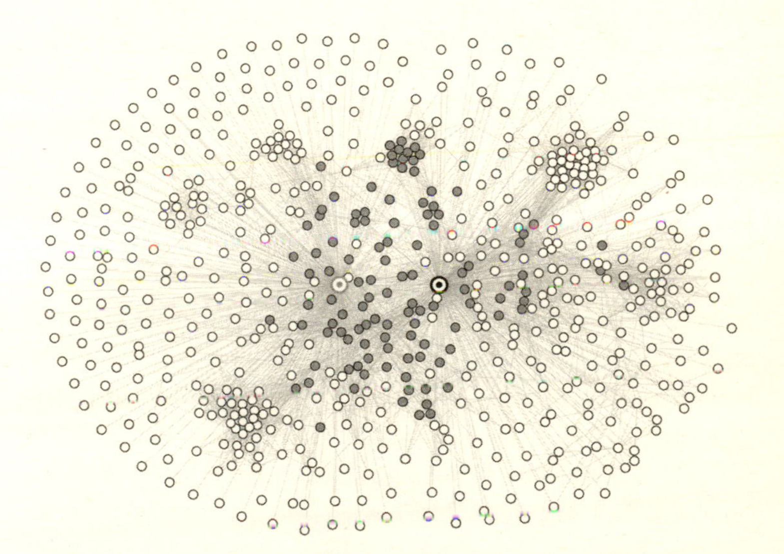

Though this might seem like a dry abstraction of a couple's life together, a mutual plot like this tells you a tremendous amount about the bond between the two people it's built around. From just the plot, the image alone, we can calculate that Reshma and I are much less likely than other couples to break up.

Network analysis, the study of dots and lines just like the patterns above, has been a science for almost three hundred years, and you can see something of the rise of data (from trickle to cataclysm) in its progress. The first network problem was a kind of rustic brainteaser, really an Enlightenment-era urban legend, that it was impossible to walk through the Prussian city of Königsberg by crossing each of its seven bridges once and only once. In 1735, Leonhard Euler, as geniuses will do, came along and reduced what had been a colloquial question of neighborhoods and footpaths to an abstraction of dots and lines (formally: nodes and edges), and in doing so, he proved with rigor that the legend was true. He expressed the town as a network, and a discipline was founded.

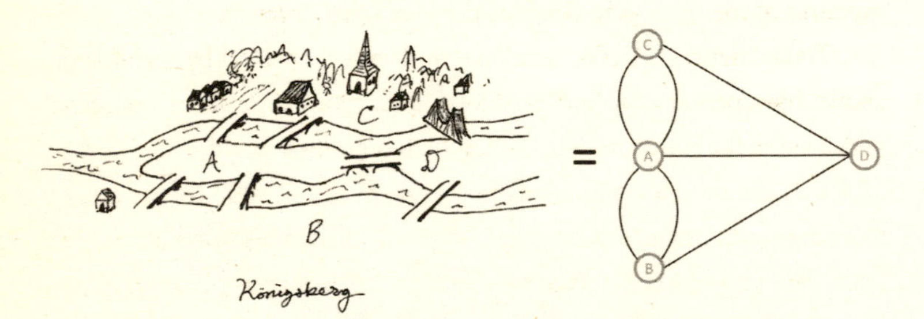

Euler's insight was that because you're only supposed to cross each bridge once, to enter a new neighborhood you need a *pair* of bridges—one to get you in, another to get you out. So the solution is as simple as looking at the network plot and asking whether each point along your path, other than your beginning and end, has an even number of lines (a pair of bridges) attached. In Königsberg, none of them do, so the problem was solved. That from such homely origins can come an enduring and flourishing science, one that's only now finding its full expression, is, I think, the best possible case for the human spirit.* Euler's concept of nodes and edges, which at first unraveled nothing more than a day's walk, has since helped us understand disease and its vectors, trucks and

* Evidence against: of the seven bridges so famous in Euler's time, four have since been destroyed. Two by bombs and two by a superhighway.

their routes, genes and their bindings, and of course, people and their relationships. And in just the last few decades, network theory's application to these last have exploded—because the networks themselves have exploded.

Forty years ago, Stanley Milgram was mailing out parcels (kits with instructions and postage-paid envelopes) to a hundred people in Omaha, working on his "six degrees of separation," hoping maybe a few dozen adventuresome souls would participate. His quaint methods—ingenious though they were—would give him the famous theory, but not quite its proof. In 2011, the unprecedented and overwhelming scale of Facebook allowed us to see that he was indeed right: 99.6 percent of the 721 million accounts at the time were connected by six steps or fewer.

Today, network theory, working on data sets enabled by technology, shows how people can find new jobs, sort information from nonsense, and even make better movies. When they built their headquarters, Pixar famously put the only bathrooms in the building inside the central atrium to force interdepartmental small talk, knowing that innovation often comes from the serendipitous collision of ideas. Theirs was an application of "the strength of weak ties," a concept postulated in the 1970s with samples in the dozens, but since amplified on new, robust network data: it tells us that it's the people you don't know very well in your life who help ideas, especially new ones, spread.*

Another long-held idea in network theory is "embeddedness." One of its expressions is the amount of overlap in a pair of social graphs—Reshma's and my embeddedness is simply how large the gray-dot portion of our graph is compared with the whole. Research using a variety of sources (e-mail, IM, telephone) has shown that the more mutual friends two people share, the stronger their relationship. More connections imply more time together, more common interests, and more stability. But unlike, say, telephone records, or even e-mail, online social networks attach rich data to a graph's edges and nodes (not unlike how dating sites have

* The original paper has been cited more than 20,000 times.

taken the timeless ritual of courtship and added age and beauty as variables to study) and of course Facebook is the richest such network ever created. The effects of that richness are just being felt.

Social-graph analysis began as, and largely remains, a matter of "who knows who." The scope of Facebook data—you can go many degrees deep with practically no added effort—is starting to turn that on its head. For relationships, and romantic relationships specifically, this data has recently enabled a new, powerful measure of how strong a bond between two people is. It turns out your lives should not just be intertwined but intertwined in a specific way. And, rare among network analysis metrics, who *doesn't* know who is the important quantity.

Two scientists, Lars Backstrom and Jon Kleinberg, working through 1.3 million couples from Facebook, established the idea in a 2013 paper. Their measure was based on counting the number of times a person and her spouse functioned as the bridge between *disjointed* parts of their network as a couple. Here's what I mean: the graph on the left below is a hunky-dory scene, more or less everybody knows one another; it is very highly embedded. But the stronger marriage is on the right. There, the couple, A and B, are the sole connectors for the two cliques around them:

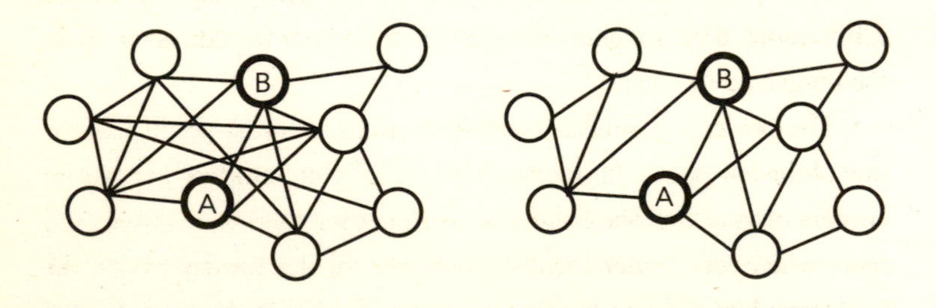

This probably feels a little strange—why would you want your network to be more fractious but for you and your spouse? But like the best ideas, it plays out intuitively in real life. For example, going back to my own story, Reshma's cousin Sheel is highly embedded in her life. The two of them grew up together, and he, like she does, has connections to virtually every member of their large extended family, including many people I don't even know. They've known each other their entire lives, whereas Reshma and I have been married for only seven years. Sheel and Reshma's relationship as a central pair would function much like my left-hand example above. However, Sheel doesn't know Reshma's coworkers. He doesn't know the members of Reshma's dance troupe. He doesn't know Reshma's friends from college. I know them all, and what's more, I am the only other person in her life these three distinct groups have in common, at least directly. For these groups, we embody the ideal on the right. It's worth noting that if, for example, Reshma and I worked together, or she didn't dance, or we went to the same college, we could not play the role we do in each other's networks.

Backstrom and Kleinberg call the level to which a relationship fulfills this ideal its "dispersion" because it shows how disconnected your graph would be without you—that is, how utterly your social circle would fly to the winds if you and your spouse were somehow ripped from the center (by, say, having a second child). I prefer "assimilation" because I think that better captures the upshot: assimilated people have a unique role *as a couple* within their mutual network. Highly assimilated couples function—the two people together—as the bond between many otherwise unconnected cliques. They are the special glue in a given spread of dots, and furthermore, they're a glue like epoxy: it takes both ingredients to make the thing hold together.

The power of assimilation comes from the fact that your spouse is one of the few people (if not the only person) you introduce into the far corners of your life. She is there at work parties, there at reunions, and there when your gamer friends come over for that all-day Magic: the Gathering blowout you look forward to all year. (Or she's not there, if she can help it, but you get the idea.) Meanwhile, these coworkers, these classmates, and these gamers, though all densely intraconnected groups themselves, are unrelated to one another but for you and your spouse.

And here's why it matters: For married people on Facebook, their spouse is the most assimilated member of their network an astounding 75 percent of the time. And, even more important for assimilation as a metric of relationship strength, the young couples for whom that's *not* the case are 50 percent more likely to break up. In the most stable relationships, the two people play this unique role in each other's lives. Considering alternate graphs of a nonassimilated couple, it makes a certain sense why—in an overly embedded one, like the left-hand example before, you and your spouse end up competing with everyone else for time and attention. There's too much leveling, no specialness. Too many girls' nights. Or in a cliquey network without assimilation, "leading separate lives" can very quickly become "leading secret lives," which might look something like this:

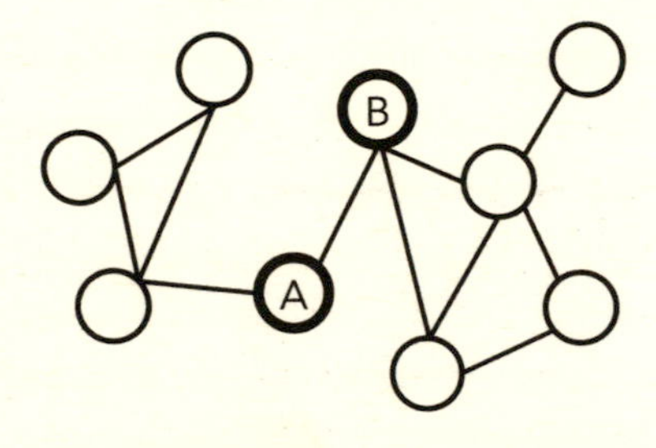

Against assimilation, Backstrom and Kleinberg tested many other ways to evaluate a relationship, and there was one detail in their paper, presented almost as an aside, that I found particularly wry. Early on, the best predictor of a relationship doesn't depend on the couple's social graph at all; for the first year or so of dating, the optimal method is how often they view each other's profile. Only over time, as the page views go down and their mutual network fills out, does assimilation come to dominate the calculus. In other words, the curiosity, discovery, and (visual) stimulation of falling for someone is eventually replaced by the graph-theory equivalent of nesting.

5.

There's No Success Like Failure

There's a great Tumblr called "Clients from Hell," where anyone can submit their service-industry horror stories. There are all kinds of cluelessness and oblivion on display, and new posts go up every few hours. Here's a typical submission, from someone doing a photo spread:

CLIENT: Can we have a heading on the photo as well?

DESIGNER: Well, it already has a caption.

CLIENT: If the reader misses the caption, then they will still see the heading.

DESIGNER: It would be quite unusual to have both a heading and a caption on a photo.

CLIENT: That makes sense. Just put a heading next to the caption, then.

My favorite client quote on the site right now is: "I don't like the dinosaur in this graphic. It looks too fake. Use a real photo of a dinosaur instead." The blog mostly gets submissions from graphic designers, but Clients from Hell's popularity speaks to a universal truth. People hate their customers.

I don't mean hate on an individual level but, en masse, customers, like any rabble, are to be feared. Anyone who tells you otherwise, from the cupcake-shop owner down the street to the CEO in the boardroom, is lying. Part of it is the ". . . is always right" thing—nobody likes a person with that much power. But by far the biggest cause of frustration is that people don't understand and can't articulate what they actually need. As Steve Jobs said, "People don't know what they want until you show it to them." What he didn't say is that showing them, especially in tech, means playing a game of Pin the Tail on the Donkey with several million people shouting advice.

If you are, say, a car company and people don't like some part of your product, they mostly tell you indirectly, by not buying it. There's historically been no open channel between Ford and the folks who want the cup holders to be green or who think it would be better if the steering wheel were a square, because, you know, most turns are 90 degrees. That's why traditional companies spend so much on market research—they have to stay way ahead of these kinds of things, because by the time a company like Ford would naturally hear about a problem, via Accounts Receivable, it's way too late.

A website is different: if people have a cockamamie idea, someone at the company is just an e-mail away. And if people don't use something, the site notices immediately. Measurements are tracked in real time, down to the finest grain, everywhere. Whenever you see something new on your favorite site—Google, Facebook, LinkedIn, YouTube, or anywhere—and you click it, know that someone, probably wearing headphones and eating Doritos, just saw a little counter go up by 1. That's when the richness of data can drive a person crazy: one of Google's best designers, the person who in fact built their visual design team, Douglas Bowman, eventually quit because the process had become too microscopic. For one button, the company couldn't decide between two shades of blue, so they launched all forty-one shades in between to see which performed better. *Know thyself:* It was etched into a footstone of the Temple of Apollo at Delphi. But like the rest of the best wisdom that time has to offer, it goes right out the window as soon as anyone turns on a computer.

Not knowing what customers need from a car, or even from a particular website interface—those are matters for a business school or a design workshop. It's when people don't understand their own hearts that I get interested. People saying one thing and doing another is pretty much par for the course in social science, but I had a rare opportunity to see people *acting* in two contradictory ways. And it all happened because *I* didn't know what they wanted either.

∞

On January 15, 2013, OkCupid declared "Love Is Blind Day" and removed everyone's profile photos from the site for a few hours. The idea was to do something different and get a little attention for a new service we were launching at the same time. The programmers "flipped the switch" at nine a.m.:

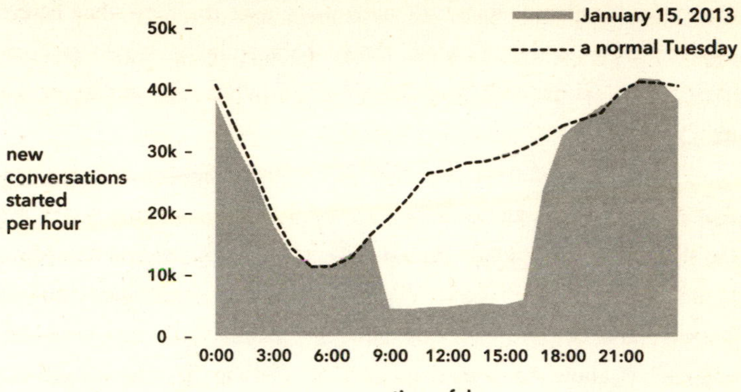

It was a bona fide pit of despair—rare in the wild! The new service OkCupid was trying to promote was a mobile app called Crazy Blind Date. With a couple taps on the screen, it would pair you with a person and select a place nearby and a time in the near future for the two of you to meet. The app provided an interface to let both parties confirm, but there was no way for anyone to directly communicate before the date. The only information it gave you about the other person was a first name and a scrambled thumbnail, like the one below. You were just supposed to show up and hope for the best.

You've probably already noticed that I'm speaking of Crazy Blind Date in the past tense. Even after a quarter million downloads, it failed, because in the end people insist on seeing what they're getting into. The app was one of those ideas that looks great on a whiteboard and miserable in the full color of creation—it was like one long "Love Is Blind Day," and with no way to flip the switch back to normal. A few months after launch, we shut the service down, but be-

a CBD-style scramble of a stock photo

fore Crazy Blind Date went off to the great app store in the sky (little-known fact: there are no bugs in heaven, just sweet features), about 10,000 people used it to share a beer or a cup of coffee with someone they'd never seen or spoken to before.

From these intrepid few, the app bequeathed the world a rare data set. Crazy Blind Date recorded not only the fact that dater A and dater B met in person but also their opinions of each other. After each completed date, like a nosy roommate, the app asked how it went. Because most of the users also had OkCupid accounts, we were able to cross-reference this data with all kinds of demographic details. We suddenly had in-person records to combine with our massive collection of digital interactions. When you merge the two sources you find something remarkable: the two people's looks had almost no effect on whether they had a good time. No matter which person was better-looking or by how much—even in cases where one blind-dater was a knockout and the other rather homely—the percent of people giving the dates a positive rating was constant. Attractiveness didn't matter. This data, from real dates, turned everything I'd seen in ten years of running a dating site on its head.

Here are the numbers for men. I've expressed attractiveness at right as the *relative difference* in a couple's individual ratings, rather than as absolutes. I did this to capture the fact that a person's happiness at finding himself across the table from, say, a "6" is highly dependent on his own looks. If he's a "1," he might be thrilled with that arrangement—it means he's dating up. A "10" would feel differently. I've included the counts of dates as the bars to show that the balance in attractiveness between the men and women going on the dates was about what you'd expect if they were randomly paired. There was no evidence of people gaming the system by, say, somehow unscrambling the pictures beforehand or showing up to the date venue and then leaving on the sly when their blind date arrived and didn't pass muster. The satisfaction numbers (for males) are the percentages in gray:

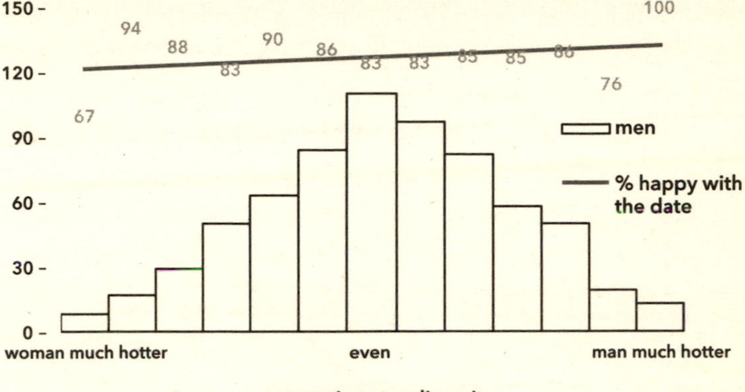

And following is the same data for women:

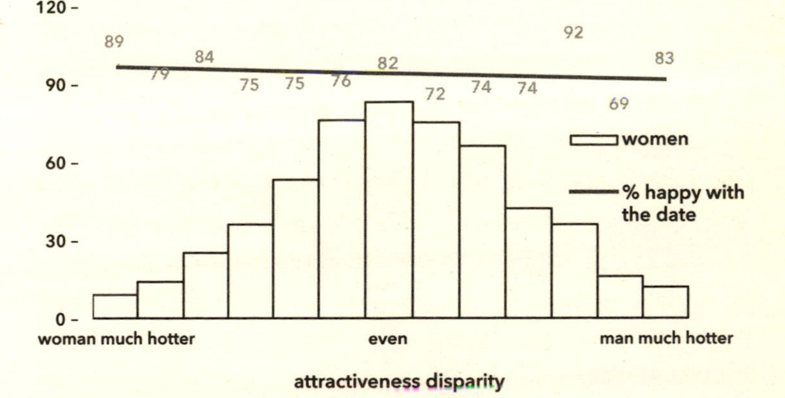

Through both Crazy Blind Date data sets, people just didn't seem to care that much about the other person's physical appearance. Women had a good time 75 percent of the time, men 85 percent. The rest of the variation is basically noise. That indifference to looks is just about the opposite of what you see in the OkCupid data. For example, I've plotted the in-person satisfaction data on the previous page (the numbers in gray) alongside those same women's *reply rates* to messages online. To make it easier to compare them, the lines show change against the average of their respective quantities:

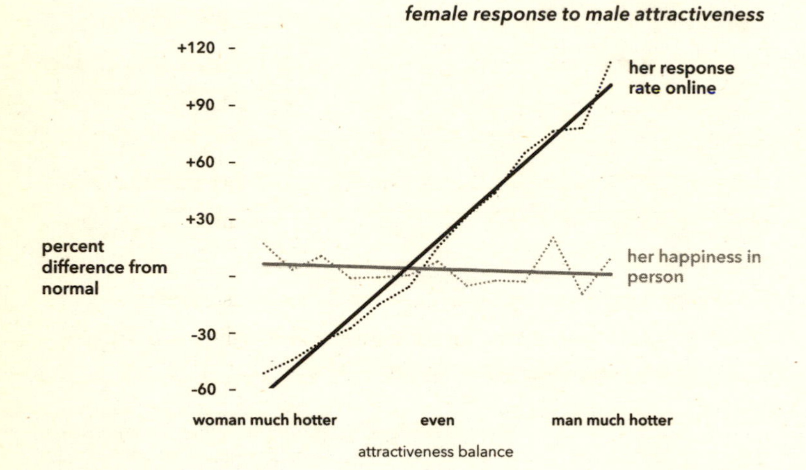

The male comparison chart is very similar to this one, and, to be clear, the data underpinning the two lines above is from the *same set of people*. The black line is their OkCupid experience, the gray from Crazy Blind Date. In short, people appear to be heavily preselecting online for something that, once they sit down in person, doesn't seem important to them.

That kind of superficial preselection is everywhere. In fact, there's a lot of money to be made off it. You know what the difference between Tylenol and Kroger's store-brand acetaminophen is? The box. Unless you take medicine like a king snake and plan to just swallow the package whole, there's really no reason to pay twice as much for the "name" molecules, whose properties are determined by immutable chemical law. And yet, I have a big red Tylenol bottle on my dresser.

We of course pay the most attention to labels when they're attached to people. In terms of superficial compatibility, self-described Democrats and Republicans get along the least of all major groups on OkCupid—worse even than Protestants and Atheists. I know this through the many match questions the site asks: they cover pretty much everything, and the average user answers about three hundred of them. The site lets you decide the importance of each question you answer, and you can pinpoint the answers that you would (and would not) accept from a potential match. Despite all this control, in the political case, the system breaks down. When you look beyond the labels, at who actually messages whom, and who replies (and therefore who ends up going on actual dates), it's *caring* about politics, one way or the other, that is actually more important to mutual compatibility than the details of any particular belief. We confirmed this in a summer-long experiment in 2011.

People tend to run wild with those match questions, marking all kinds of stuff as "mandatory," in essence putting a checklist to the world: I'm looking for a dog-loving, agnostic, nonsmoking liberal who's never had kids—and who's good in bed, of course. But very humble questions like *Do you like scary movies?* and *Have you ever traveled alone to another country?* have amazing predictive power.* If you're ever stumped on what to ask someone on a first date, try those. In about three-quarters of the long-term couples OkCupid has ever brought together, both people have answered them the same way, either both "yes" or both "no." People tend to overemphasize the big, splashy things: faith, politics, and certainly looks, but they don't matter nearly as much as everyone thinks. Sometimes they don't matter at all.

Fiasco though it was, Love Is Blind Day gave us a visceral example of what people do in the absence of information. In hiding pictures but changing nothing else, we created a real-time experiment to set against the site's usual activity. For seven hours our users acted without the very thing our previous data had indicated was the single most important piece of knowledge OkCupid could offer: what everyone else looked like.

Some of the upshot was predictable. People sent messages without

---

* The "scary movies" question originally asked about "horror movies." When we realized what a good question it was, we moved it much higher in the queue and changed "horror" to "scary" as an editorial decision. We felt that second word was less . . . scary . . . to new users, who would now be seeing the question much more often. We often tweak questions like this.

the typical biases, or racial and attractiveness skews. What a user couldn't see, he couldn't judge. But of the 30,333 messages sent blindly, eventually 8,912 got replies, a rate about 40 percent higher than usual. And in the dark, for those who were there, something astounding happened. Twenty-four percent of the pairs of people talking when the photos were hidden had exchanged contact info before pictures were turned back on. That was in only the seven-hour window of Love Is Blind Day. The expected number in that amount of time is barely half that. So not only were people writing messages that were far more likely to get replies, they were giving out phone numbers and e-mail addresses at a higher rate—to people they'd never even seen.

For the couples who began talking and were still getting to know each other when we restored photos at four p.m., however, the day had a reverse effect. The two people had been in the dark, then suddenly the lights came on, and, in the data, you can actually see them spook. Threads straddling the moment we flipped the switch lasted an average of 4.4 more messages. When you compare them against a control data set, they should've lasted 5.6. Eventual contact-info exchanges in those "lights on" threads were down by a similar amount.

Dating sites are designed to give people the tools and the information to get whatever they want out of being single—casual sex, a few fun dates, a partner, a marriage . . . anything. Stuff like height, political views, photos, essays, all of it is right there, easily sortable, easily searchable. It's there to help people make judgments and fulfill their desires, and as fascinating as those judgments and desires may be to pick apart, there's a side of it that I think does love a disservice. People make choices from the information we provide because they *can*, not because they necessarily should.

I can't help think of the many people getting turned down because of some perceived "deal-breaker" that actually no one cares about and wonder if the Internet has changed romance in the way it's changed so much else—and for the same reason. If I may channel my inner anti-Jagger: Online, you *can* always get what you want. But what you need, that's a much harder thing to find.

PART 2

What Pulls Us Apart

The Confounding Factor

If you stand on the southwest corner of Fifty-Eighth and Fifth with a clipboard and do a little people-watching, you can very quickly conclude that most New Yorkers are beautiful, thin, and above all, rich. Every thread, every grommet, every crease shines with money. Of course, many New Yorkers *are* rich, but that's not the whole story here. You're standing outside Bergdorf Goodman, and that's a confounding factor.

This is a technical term for something you haven't accounted for in your analysis but that nonetheless affects its results. Making sure you're not perched in some bitwise version of the Upper East Side is one of the most time- and thought-consuming parts of working with digital data. When you have seemingly every variable and every possibility available for analysis and speculation, your research is free to travel wherever your curiosity leads. But true to the cliché, that freedom requires eternal vigilance.

And here's where I have an admission to make. So far in these pages, wherever you've seen the data of a person-to-person opinion, in the votes, in the date results from Crazy Blind Date, the charts, the tables—in every ratio, in every total—whenever one user was judging another, both people involved were white. I had to make it that way, because when you're looking at how two American strangers behave in a romantic context, race is the ultimate confounding factor. And to make sure whatever I wanted to say about attraction or sex spoke to those ideas alone, I needed to cut it from the discussion.

As an American, the reflex to sweep race under the rug is inborn, so in a way, though the numbers forced my hand, I was just doing what came naturally. And even apart from our nation's peculiar relationship with the topic, a long history of tokenism and sorry pseudoscience makes any quantitative analysis of race especially fraught. That's not to say we don't have good numbers. There are plenty of them, of a certain type—if my preferred data is person-to-person, then I think of this other as person-to-thing: one group or another versus unemployment rates, the SAT, the criminal justice system, cancer . . . As much as research like this has helped us pinpoint and (occasionally) address inequality, there's something incomplete about it. You lose the human who is doing (or not doing) the hiring, the teaching, the police work, the preventative care; you lose the people who created the outcomes that all these studies purport to measure. So what you end up with is conclusions like this: *Black Defendants Are at Least 30 Percent More Likely to Be Imprisoned Than White Defendants for the Same Crime*. The headline's passive voice says it all. Who's miscarrying the justice here? Syntactically, no one. Practically, I have a good guess. But it is a rare study indeed that looks beyond the institutions, to the fundamental "us versus them" binary of race relations.

Behind every bit in my data, there are *two* people, the actor and the acted upon, and the fact that we can see each as equals in the process is new. If there is a "-clysm" part of the whole data thing, if this book's title isn't more than just a semi-clever pun or accident of the alphabet—then this is it. It allows us to see the full human experience at once, not just whatever side we happen to be paying attention to at a given time.

Before the advent of data like ours, one of the most quantified arenas in public life was sports. There you have real-time numbers on every conceivable interaction, and you have the data on an individual level, to be sliced and recombined at will. Perhaps it's surprising, then, that sports is where the discussion of race is *least* analytic. The "black quarterback" controversy that stretched for the first ten years or so of this millennium is the perfect example. For years there was a regular news cycle: an African American quarterback would go early in the draft or start a high-profile game, and someone would inevitably imply that blacks can't succeed at the position in the NFL. The usual reason given was that they lacked the intelligence. There would be backlash, discussion, and plenty of argument that this was nothing more than mean-spirited stereotyping. But amidst all the commentary and outcry, and outcry against the outcry, in the 97,000 results that Google returns for "black quarterback," I found only one article that actually calculates the quarterback ratings of blacks and whites, which turn out to be the same down to the second decimal: 81.55. In a genre so stats-obsessed, where platoons of number crunchers calculate Johnny Placekicker's 54 percent success rate on field goal attempts over 50 yards in road games decided by 7 points or less against AFC opponents, you'd think that statistically comparing black and white quarterbacks would've been everyone's first instinct. Instead, there was, and generally is around race, an eerie numerical silence. You find in its place rhetoric and appeals to anecdote. But a "debate" done in this style just leaves everyone believing they're right, when, in fact, for all the words expended, a single number—81.55—can clearly show that one side is wrong. The article that did the rating calculation had 0 tweets and 0 Facebook likes, by the way, and it wasn't posted on some obscure blog; it appeared on *The Big Lead,* which is owned by *USA Today*. You often get the feeling that people just don't *want* to know.

Where in situations like this we might seem to lack the *will* to examine race through a statistical lens, in many other arenas we have simply lacked the data. Most aspects of life haven't been as obsessively quantified as football. That is changing rapidly.

On OkCupid, one of the easiest ways to compare a black person and a white person (or any two people of any race) is to look at their "match percentage." That's the site's term for compatibility. It asks users a bunch of questions, they give answers, and an algorithm predicts how well any two of them would get along over, say, a beer or dinner. Unlike other features on OkCupid, there is no visual component to match percentage. The number between two people only reflects what you might call their inner selves—everything about what they believe, need, and want, even what they think is funny, but nothing about what they look like. Judging by just this compatibility measure, the four largest racial groups on OkCupid—Asian, black, Latino, and white—all get along about the same.* In fact, race has less effect on match percentage than religion, politics, or education. Among the details that users believe are important, the closest comparison

* Of course, not every person on OkCupid puts themselves in one of these neat categories. However, to simplify and focus the discussion, we'll limit our analysis to users who have selected one of the four.

to race is Zodiac sign, which has no effect at all. To a computer not acculturated to the categories, "Asian" and "black" and "white" could just as easily be "Aries" and "Virgo" and "Capricorn."

But this racial neutrality is only in theory; things change once the users' own opinions, and not just the color-blind workings of an algorithm, come into play. Given the full profile, with the photo dominating the page, this is how OkCupid's users rate each other by race:

average ratings given from men to women on OkCupid

		her race			
		Asian	**black**	**Latina**	**white**
his race	**Asian**	**3.16**	**1.97**	**2.74**	**2.85**
	black	**3.40**	**3.31**	**3.43**	**3.23**
	Latino	**3.13**	**2.24**	**3.37**	**3.19**
	white	**2.91**	**2.04**	**2.82**	**2.98**

I've given the raw data above, unadorned, because by now you're at least a little familiar with OkCupid's 1- to 5-star system. But to make the trends easier to see, I'm going to take that same matrix and "normalize" each row. In the table below, each entry is the percentage difference (+/-) from the average (the "normal") in the row. It's the same information, just phrased a bit differently. Think of the normalized number as the men's relative preference for women. For example, as you can see, Asian men think Asian women are 18 percent better-looking than the average, while black men think they're just 2 percent better. And so on:

normalized rating from men to women on OkCupid

		her race			
		Asian	**black**	**Latina**	**white**
his race	**Asian**	**+18%**	-27%	**+2%**	**+7%**
	black	**+2%**	-1%	**+3%**	-4%
	Latino	**+5%**	-25%	**+13%**	**+7%**
	white	**+8%**	-24%	**+5%**	**+11%**

I'll soon move beyond OkCupid, and when I present similar matrices later, I'll go directly to the normalized scores. But for now, the two essential patterns of male-to-female attraction are plain: men tend to like women of their own race. Far more than that, though, they don't like black women. Message data is highly correlated with these ratings, so they follow the pattern as well.*

Just to show that these voting trends aren't being thrown off by some obscure statistical artifact, I've put the raw per capita vote numbers in what's called a box plot—it tells you where the bulk of a data set lies. You see below that the central mass of black women is rated almost entirely below the other three ethnicities', and the black women's upper extreme is about at the midline of the other three:

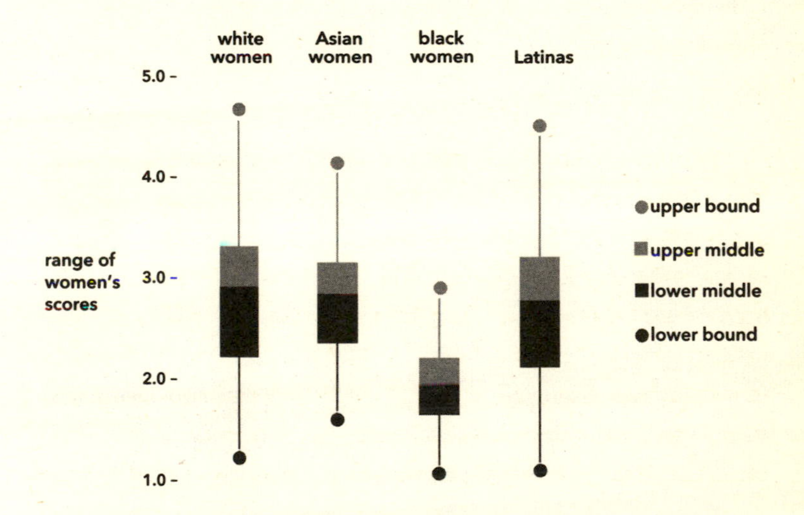

Mathematically, this is a complete discount—being black basically costs you about three-quarters of a star in your rating, even if you're at the top. Further, when you do this analysis in reverse, and look at the people actually casting the votes, you see a similar wholesale pattern. The broad majority of

* Black women get roughly 75 percent of the number of first messages that other women do. Their messages are replied to about 75 percent as often.

non-black men apply that three-quarters reduction to black women. There is no cadre of racists single-handedly bringing everything down.

However startling this may be, it only reflects one data set, the thoughts of one group of people. So here's a good place to pause for a second and answer a question you might have been asking earlier, given how much I've relied on OkCupid's data so far in this book: *Who are these people?*

In the most superficial way, OkCupid's members reflect the general composition of Internet users, with of course the caveat that (almost) everyone on the site is single. The site's users are younger than the national average (OkCupid's median age is twenty-nine), and they tend to be less religious. The racial composition is about what you'd expect. Here are our numbers compared with the generic "American Internet User" breakout from Quantcast, the major online audience measurement firm—it's like Nielsen for the net.

racial composition

	OkCupid users	American Internet users
Asian	**6%**	**4%**
black	**7%**	**9%**
Latino	**8%**	**9%**
white	**80%**	**78%**

Going one demographic level deeper, OkCupid users are, if anything, more urban, more educated, and more progressive than the nation at large. The site's biggest markets by far are places like New York, San Francisco, Los Angeles, Boston, and Seattle. Eighty-five percent of the users have gone to college. Self-described liberals outnumber self-described conservatives more than two to one. There is a broad, site-wide ethos of open-mindedness. And an unintentionally hilarious 84 percent of users answer this match question...

Would you consider dating someone who has vocalized a strong negative bias toward a certain race of people?

in the absolute negative (choosing "No" over "Yes" and "It Depends"). In light of the previous data, that means 84 percent of people on OkCupid would not consider dating someone on OkCupid.

Essentially anything that, in theory, would make a group of people "less racist," that's what OkCupid users are. I point this out to people, who, like me, lead nice lives in large, diverse cities; who think of their opinions and tastes as nothing if not enlightened; who unwind at night with a glass of wine and a Facebook dose or two of progressive righteousness: When I show here that black women and later, black men, get short shrift, and that adding whiteness to a user's identity makes him or her more attractive, I'm not describing some Ozark fever dream. I'm describing our world, mine and yours. If you're reading a popular science book about Big Data and all its portents, rest assured the data in it is you.

But look one more time at the match question above, which was written by one of OkCupid's users and has been answered close to two million times: "vocalized" is an odd word. Get rid of it, and it still more or less reads "Would you date a racist?," which I once assumed was the question's real intent. The writer, however, understood the subtleties of the data set before I did. On a dating site you can act on impulses that you might otherwise keep quiet. On some level, the users come to judge and be judged by others, and each person joins the site free of the context of their everyday life. The site doesn't connect you to your family. Nothing gets posted to your friends' timelines. The game is: it shows you people, and you like them or you don't; you talk to them or you don't. There's nothing else to it. In a digital world that's otherwise compulsively networked, there's an old-school solitude to online dating. Your experience is just you and the people you choose to be with; and what you do is secret. Often the very fact that you have an account—let alone what you do with it—is unknown to your friends. So people can act on attitudes and desires relatively free from social pressure.

In the layperson's mind, Facebook, "the social network," is the sine qua non of online data sources. And it's easy to see why: Facebook is huge and pervasive, and a sample of their users is pretty much a sample of people worldwide who have Internet access—in other words, you can easily get a representative corpus for whatever you want. And they have such robust and diverse data: they know who you went to high school with, what song you just listened to on Spotify, where your parents live, and so on.

But as often as it is an asset, that richness can be a liability. You rarely meet a stranger on Facebook. The site is, by design, people you already know and whom you've already made up your mind about—they're your friends, after all. Facebook's data on race is the embodiment of the "But I have black friends" solipsism you often hear. How you treat your friends is, by definition, the exception to how you treat the rest of humanity. And you and your friends' relationships were formed outside of the network first.

Moreover, people become inhibited when their friends are watching. This fishbowl aspect is why the first step of most dating apps on Facebook is to get you off Facebook—your existence there is fully chaperoned. Long ago, we tried "social" features on OkCupid, and they bombed, as did similar features when Match.com gave them a go. For whatever reason, people don't want their network along for online dating. The desire for solitude comes from the same place, I imagine, as the claustrophobia that would grip most of us if, on a promising first date at some restaurant, two old friends posted up at a nearby table. This is to take nothing away from the business or the community Facebook has created, but the "real life" relationships that both undergird and overarch the site give a different power to their data. When you want to look at something like race, where, at least among decent people, there's pressure to behave a certain way in public, dating sites provide a uniquely powerful data set: everyone's a stranger, alone, and there to tell you who they like and who they don't.*

So then let's put OkCupid's data up against data from other dating

* Now, of course, dating sites are far from a perfect *general* source. As we both know, almost every user is single, and that has consequences. Using our data, if I were to sit here and research, say, spending habits, and thus conclude that the average American man spends all his disposable income on restaurants and movie tickets, I'd be making a fool of myself. A claim like this, oblivious to the special nature of my source, would be absurd.

sites and see what shakes out. Looking at numbers made by other users, acting through other interfaces, gives us a much better sense of the real pattern. And that's what we see below—this is data from OkCupid, DateHookup, and Match.com, sites that together signed up about 20 million Americans last year alone, presented side-by-side. In the particulars, the matrices vary—remember, these values reflect actions produced by different people using different software—but cutting through that difference is the same broad pattern. In terms of the "direction" of feeling, like or dislike, these matrices are very nearly identical:

		her race			
	OkC	**Asian**	**black**	**Latina**	**white**
his race	**Asian**	**+18%**	-27%	**+2%**	**+7%**
	black	**+2%**	-1%	**+3%**	-4%
	Latino	**+5%**	-25%	**+13%**	**+7%**
	white	**+8%**	-24%	**+5%**	**+11%**

Match	**Asian**	**black**	**Latina**	**white**
Asian	**+50%**	-68%	-14%	**+31%**
black	**+9%**	-13%	**+8%**	-3%
Latino	**+4%**	-67%	**+33%**	**+29%**
white	**+13%**	-68%	**+8%**	**+47%**

DH	**Asian**	**black**	**Latina**	**white**
Asian	**+11%**	-24%	**+9%**	**+4%**
black	**+7%**	-9%	**+9%**	-7%
Latino	**+12%**	-27%	**+10%**	**+6%**
white	**+18%**	-30%	**+6%**	**+5%**

Match.com, you probably know. It's been the most popular dating site in the United States for almost two decades. They buy tons of advertising on national television and, as a result, have exactly the broad "all-American" demographics you'd expect. DateHookup is a free site of several million members that is very popular among casual daters; its user base is just under 20 percent black and 13 percent Latino. It's the most diverse of the three sites considered here. I think of it as the Atlanta or the Houston to OkCupid's Portland and Match's Dallas. But as you see, across all three sites, for men rating women, you get the same pattern wherever you go.

The votes in the other direction, of women rating men, aren't quite as uniform from site to site, though they're still very similar:

	his race				
	OkC	**Asian**	**black**	**Latino**	**white**
her race	**Asian**	**+10%**	-20%	-8%	**+19%**
	black	-16%	**+24%**	-8%	-0%
	Latina	-19%	-11%	**+10%**	**+20%**
	white	-14%	-11%	-1%	**+25%**

Match	**Asian**	**black**	**Latino**	**white**
Asian	**+21%**	-49%	-38%	**+66%**
black	-50%	**+53%**	-6%	**+2%**
Latina	-54%	-37%	**+39%**	**+53%**
white	-49%	-39%	-2%	**+90%**

DH	**Asian**	**black**	**Latino**	**white**
Asian	-	-21%	**+9%**	**+13%**
black	**+5%**	**+13%**	-6%	-12%
Latina	-11%	-9%	**+14%**	**+7%**
white	-8%	-16%	**+4%**	**+19%**

These matrices show two negative trends, and two positive. Blacks are again unappreciated by non-black users, but Asian men have joined them in the deep gray. On the positive side, women clearly prefer men of their own race—they're more "race-loyal" than men—but they also express a preference for white men.

Another way to dig into racial hierarchies is open to us on OkCupid, and it reinforces this "white preference." Because the users are able to select more than one ethnic identity, we can study racial blends in an almost laboratory-like way. For example, we have men who check "Asian" as their ethnicity. We also have men who check both "Asian" and "white." Comparing the two groups gives us some sense of what adding "whiteness" gets a person. It turns out: quite a bit. When you add white, ratings go up, across the board. I've just spilled out the complete data here. It's a big, messy table, but it's worth exploring.

Down the right-hand column you see the improvement in scores created by whiteness in a person's racial makeup. The biggest takeaway is that the racial discount applied to black men and women and Asian men in the tables above is significantly undone here. It's the reverse of the old "one-drop" rule.

Unfortunately, there aren't enough people who select "black" and "Latino" or "Asian" and "black" to fully flesh out this alchemy, but it's an intriguing glimpse at how we view the ethnic spectrum:

		her race		
	men rating women	**Latina**	**Latina + white**	**% change**
	Asian	2.7	2.8	+4
	black	3.4	3.4	-2
	Latino	3.4	3.4	+1
	white	2.8	3.0	+7
		black	**black + white**	
	Asian	2.0	2.3	+19
	black	3.3	3.5	+5
race	**Latino**	2.2	2.9	+28
	white	2.0	2.5	+24
		Asian	**Asian + white**	
	Asian	3.2	3.0	-5
	black	3.4	3.6	+5
	Latino	3.1	3.3	+5
	white	2.9	3.0	+2

	women rating men	his race: Latino	Latino + white	% change
her race	Asian	1.7	1.8	+7
	black	2.0	2.4	+18
	Latina	2.1	2.2	+8
	white	1.8	2.1	+15
		black	black + white	
	Asian	1.5	1.6	+6
	black	2.7	2.6	-4
	Latina	1.7	1.9	+17
	white	1.6	2.0	+26
		Asian	Asian + white	
	Asian	2.0	2.1	+4
	black	1.8	2.7	+48
	Latina	1.5	2.2	+44
	white	1.5	2.0	+32

Now, this is all taken from ratings on a dating website, but dating data is essentially data of the first impression, of the first blush—the users need to get to know each other, at least a little, before they're going to want to kiss—and it's in that same basic spirit that any pair of people come together: *Well, what am I looking at? Who do I see?* The data measures the frisson of meeting someone new: that burst of judgment and instinct and chemistry that determines whether you like a person or not, before you even really know much about them. Here are a few OkCupid users putting it in their own words:

> Then one day, I think I was looking through my daily matches and there he was. I instantly clicked on his profile . . . something about him, just made me smile.
>
> —Bella, on Patrick

> Well, it all began when one day I am looking through my matches and see this girl that I found attractive from first glance.
>
> —Dan, on Jenn

But if there is love at first sight, there is dislike at first sight too, right? And is it not that same frisson of attraction, but in reverse, when someone flinches, however unconsciously, from a stranger? Here, again, someone in his own words:

> There are very few African American men who haven't had the experience of walking across the street and hearing the locks click on the doors of cars. That happens to me. . . . There are very few African Americans who haven't had the experience of getting on an elevator and a woman clutching her purse nervously and holding her breath until she had the chance to get off. That happens often.
>
> —Barack Obama, July 19, 2013

These flashes of intuition at the core of the data—extrapolations from just the smallest amount of information—pertain not just in romance, but in picking who you rent your apartment to, in deciding to approve a loan or not, and, surely, in police work, where there's often no time for anything but a flash. Even in more deliberate situations, the first impression plays the heavy. One paper asked: "Are Emily and Greg More Employable Than Lakisha and Jamal?" and got a resounding "Yes" from our nation's HR professionals. The scientists sent identical résumés, some with "black-sounding" names at the top and some with "white-sounding" ones, and found that the latter received 50 percent more responses, no matter the position or industry. And companies that say they're "Equal Opportunity Employers" discriminate as much as anyone else.

That kind of irony gets to why big studies are important, but small person-to-person measurements are essential: when you read findings like the one above, and see that Jamal doesn't get the job, it's easy to shake your head at the few racist hiring managers who've tilted the odds against him. But the data we see in this chapter shows racism isn't a problem of outliers. It is pervasive. We've seen the same patterns repeated on three different sites, with different users and different experiences: men, women, free, subscription-only, casual, serious, "urban" demographics, and more "mainstream." All told, the research set represents a large chunk of the young adults in this country, and the data uniformly shows non-blacks discount African American profiles. It's not a problem caused by a small cluster of "ugly" black users or by a small group of unreformed racists throwing off an otherwise regular pattern.

It is no longer socially acceptable to be openly racist. In response to that pressure, there is some portion of the public who have therefore slunk away: if I can't shout hate at some schoolchildren anymore, well, fine, I'll just shout it at the TV. This is not the typical American. Most of us—almost all, in fact—recognize that racism is wrong. But it is still implicit in many of the decisions we make.* Psychologists have a name for the interior patterns of belief that help a person organize information as he encounters it: *schema*. And our schema is still out of step with how most of us know the world should be. By hundreds of small, everyday actions, none of them made with racist intent or feeling, we reflect a broader culture that is, in fact, racist. As we've seen, the pattern is so woven-in that relatively recent additions to our society, Asians and Latinos, have adopted it, too.

When it comes to these patterns, the individuals are, in a way, blameless. That black people get three-quarters the affection on dating sites is practically an accident. I can't fault someone for not wanting to go on a date with someone else. There's rarely any malice in that decision. Judgments like votes are made in an instant, and are such small, seemingly meaningless, things. You browse around and maybe one face in twelve is black. And looking at that person your action at that time could go in any direction, just as it could if you were looking at a white user; you're in the flow. And so what if you don't like one particular person at one particular moment? It is everyone's right to think what they want about any individual—in fact, seeing each person as an individual in the first place,

---

* To be clear, "we" isn't rhetorical. It means me, too.

and not as a category, is a huge step in the right direction. It's just that the patterns in aggregate show that the dice, overall, are still loaded. Actually, a better metaphor from the same general category: they show that the house is still taking a rake—it's not the dealer, it's not the hand, it's not even the play, it's the rules of the game that make certain groups of people lose and others win.

Sociology professor Osagie K. Obasogie recently produced some ingenious research—he interviewed people blind from birth and found the same attitudes about race as in the sighted world. His sample was relatively small—just 106 individuals, but he found my OkCupid data in the flesh. He cites numerous examples of a young blind person being happy on a date until some "tell"—usually the feel of the hair but occasionally a whisper from a stranger—revealed that the other person was black. The date was then over.

Obasogie asserts that blind people's attitudes on race reflect a lifetime of cultural absorption, as opposed to any visual reality. From his data, it seems impossible to argue otherwise. Moreover, he observed that sex is the locus of the sharpest discord between what we're looking at and what our culture tells us we see. As he puts it to the *Boston Globe,* he was struck by the vigilance with which, even among his blind subjects, "racial boundaries get patrolled, primarily in the realm of dating." To take his metaphor one step further, a patrol protects the interior, and here dating is just the frontier of a vast cultural mass that will take decades to rearrange.

Anyhow, I'm well aware of the long and embarrassing history of "science" by white researchers conducted to "prove" the scientist's belief that white people are better. And I'm equally well aware of how data showing that, just for example, "women find white men attractive" can come across. It is not my claim that white men are unusually good-looking. Nor am I claiming that the data "proves" black people aren't attractive. In fact, OkCupid's patterns change in places outside the United States. In the UK, the site's black members get 98.9 percent of the messages white members do. In Japan, 97.8 percent. In Canada, 90 percent. Many of the black users in the former two countries, especially Japan, are Americans abroad.

Sex sometimes has nothing to do with bone structure and muscle and flesh—the flaws and boons of which all races share in equal amounts. There is culture there too, and expectation, and conditioning. That's what this data shows, and because it's person-to-person, and collected in fine detail, it can show it in a way that no other research can.

I was an exchange student in Japan for a summer in high school, and the agency officials in my host town, Utsunomiya, would occasionally collect me and the other Americans to visit a school or a factory nearby. The goal was as much for us to see the country as for it to see us. This was the early '90s, pre-Internet, and Japan, not China, was still our big economic rival. There was tension; they had bought Rockefeller Center a few years before; the yen was threatening the dollar. The name of my exchange program captured the timbre of the visit in three words: Youth for Understanding.

The name notwithstanding, I found the culture baffling. I remember even the characters' names in Street Fighter II were all wrong; Vega was called Balrog and Balrog was M. Bison. . . . I was like, *This is madness*. But they did have American television; *Baywatch* would soon be the number one show in the country. At one school they bundled us off to, we had to get up and say a few words in front of the student assembly. I rose from the floor to the podium, said something dumb, and stepped down. The next person due up was the only blonde in our little troupe, and as she stood, and I'll never forget it, there was an audible gasp. The person standing there was just a regular girl—we were sixteen and all lumpy and horrid—but a shudder went through the crowd as if Pamela Anderson were there in the flesh.

Many people have taken that shudder at face value. And for decades, phrenologists, racialists, and quacks have jumped through hoops to give that essentially cultural response a biological (and therefore immutable) basis. Nell Irvin Painter's book *The History of White People* gives an excellent overview of "race science," and in the course of it she offers up a quote from an Enlightenment-era text on the wonders of the "Caucasian" race, written, naturally, by a white man:

> The blood of Georgia is the best of the East, and perhaps in the world. I have not observed a single ugly face in that country, in either sex; but I have seen angelical ones. Nature has there lavished upon the women beauties which are not to be seen elsewhere . . . it would be impossible to point to more charming visages, or better figures, than those of the Georgians.

Johann Blumenbach was the writer here; he developed his racial theories by collecting and comparing human skulls. Scholarship, perhaps, has progressed. The subconscious is another story.

7.

The Beauty Myth in Apotheosis

I work in a universe where people identify themselves along almost every conceivable axis—as smokers and non-; as Christians and atheists; as nerds or geeks, or maybe dorks; to say nothing of black or white or Asian or gay or straight, or neither, or both. Mankind is tribes within tribes. Or, putting it more beautifully, like the Korean proverb: "Over the mountains, mountains." That's the ruggedness of their peninsula and the endless difficulty of our fractured human terrain.

Running a dating site you become aware of a subdivision that on the one hand seems frivolous but on the other is as inborn as a person's race or sexuality, and like those latter traits it's often resistant to direct analysis. On OkCupid—as on Match, as on Tinder—a prime divide, perhaps the deepest, is between the beautiful and the rest. These are our haves and have-nots, our rich, our poor, and when it comes to sexual attention, the haves reap the benefit of their inheritance just as surely as any heir, while the have-nots largely go without. Not unlike race, beauty is a card you're dealt, and it has huge repercussions.

Below I've plotted new messages received per week, by the recipient's physical attractiveness:

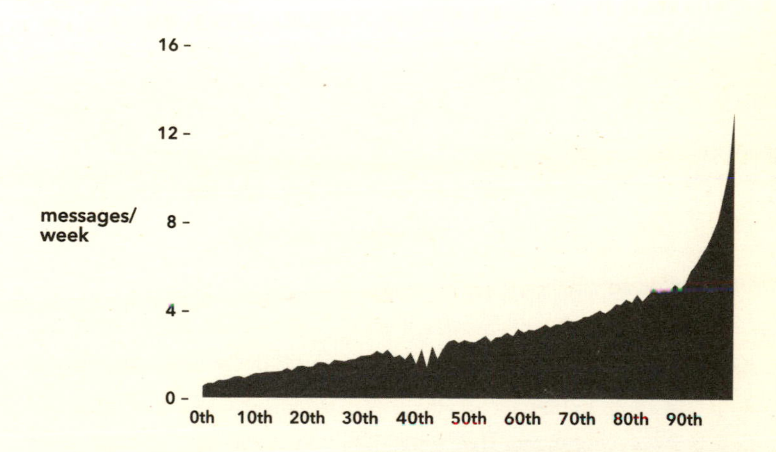

The sharp rise out at the right smashes down the rest of the curve, so its true nature is a bit obscured, but from the lowest percentile up, this is roughly an exponential function. That is, it obeys the same math seismologists use to measure the energy released by earthquakes: beauty operates on a Richter scale. In terms of its effect, there is little noticeable difference between, say, a 1.0 and 2.0—these cause tremors that vary only in degree of imperceptibility. But at the high end, a small difference has cataclysmic impact. A 9.0 is intense, but a 10.0 can rupture the world. Or launch a thousand ships.

What you definitely can't see in the chart above, because I aggregated the data to obscure it, is that men and women experience beauty unequally. Here is that OkCupid message density, split out by gender, with the aggregates as the dotted line in the middle.

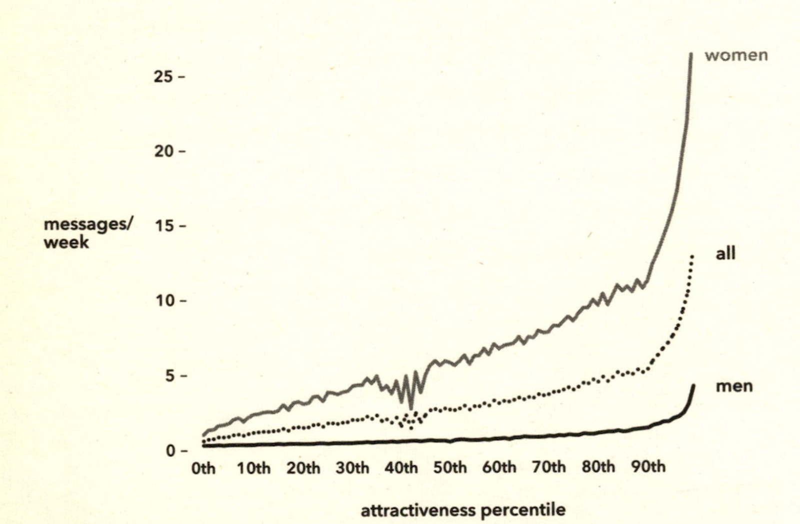

It's hard for me to convey how much attention the upper-right corner of this curve entails, short of tracking you down and screaming in your face about my hobbies. Especially in larger cities, where the message flow is 50 percent higher than even what you see above, a woman at the top of the scale has something like a term paper's worth of hey-what's-up-do-you-like-motorcycles-because-I-like-motorcycles waiting for her every time she comes to the site. A dudeclysm, if you will. However, neither beauty's effects, nor the male/female split, are confined to the sexual realm.

Here is data for interview requests on Shiftgig, a job-search site for hourly and service workers:*

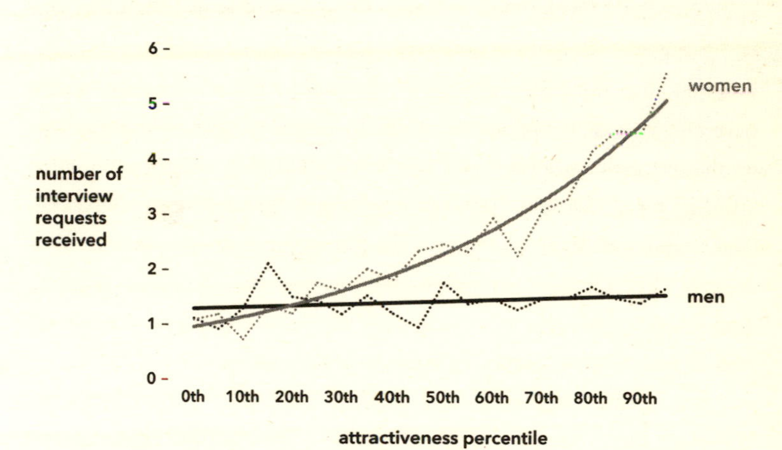

And for friend counts on Facebook:

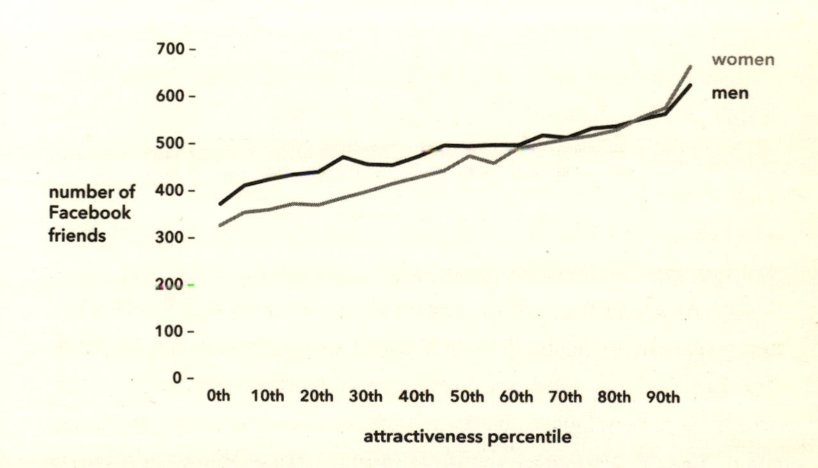

* I foreground trend lines here because the data is slightly sparser and therefore more noisy than usual. This sample is ≈5,000 people.

Success and beauty are correlated for both sexes, but you can see that the slope of the gray line is always steeper. On Facebook, every percentile of attractiveness gives a man two new friends. It gives a woman three. On Shiftgig, the curves aren't even comparable in this way. The female curve is exponential and the male is linear. Moreover, they hold whether the *hiring manager,* the person doing the interviewing, is a man or a woman. In either case, the male candidates' curves are a flat line—a man's looks have no effect on his prospects—and the female graphs are exponential. So these women are treated as if they're on OkCupid, even though they're applying for a job. Male HR reps weigh the female applicants' beauty as they would in a romantic setting—which is either depressing or very, very exciting, depending on whether you're a lawyer with a litigation practice. And female employers view it through the same (seemingly sexualized) lens, despite there (typically) being no romantic intent.

It is hardly fresh intellectual ground that beauty matters, and that it matters more for women. For example, a foundational paper of social psychology is called "What Is Beautiful Is Good." It was the first in a now long line of research to establish that good-looking people are seen as more intelligent, more competent, and more trustworthy than the rest of us. More attractive people get better jobs. They are also acquitted more often in court, and, failing that, they get lighter sentences. As Robert Sapolsky notes in the *Wall Street Journal,* two Duke neuropsychologists are working on why: "The medial orbitofrontal cortex of the brain is involved in rating both the beauty of a face and the goodness of a behavior, and the level of activity in that region during one of those tasks predicts the level during the other. In other words, the brain . . . assumes that cheekbones tell you something about minds and hearts." On a neurological level, the brain registers that ping of sexual attraction—*Ooh, she's hot*—and everything else seems to be splash damage.

To my second point, that beauty affects women in particular, Naomi Wolf's bestseller *The Beauty Myth* showed that better than I ever could. In short, my raw findings here are not new. What is new is our ability to test ideas, established ones, famous ones even, against the atomized actions of millions. That granularity gives strength and nuance to previous work, and even suggests ways to build on it.

The paper "What Is Beautiful" was based on a research sample of only 60 subjects—barely adequate to prove the effect, let alone its many facets.* But now we can go from "What Is Beautiful Is Good" to asking "How Good?" and in what contexts. In sex, beauty is very good. In friendship, it's only somewhat good, and when you're looking for a job, the effect really depends on your gender. As for Wolf's seminal work, we can confirm the truth behind her broad observation that "today's woman has become her 'beauty'"—three robust research sets agree that the correlation is strong. And, better, we can extend some of her most cogent arguments about beauty being a means of social control. Think about how the Shiftgig data changes our understanding of women's perceived workplace performance. They are evidently being sought out (and exponentially so) for a trait that has nothing to do with their ability to do a job well. Meanwhile, men have no such selection imposed. It is therefore simple probability that women's failure rate, as a whole, will be higher. And, crucially, the criteria are to blame, not the people. Imagine if men, no matter the job, were hired for their physical strength. You would, *by design,* end up with strong men facing challenges that strength has nothing to do with. In the same way, to hire women based on their looks is to (statistically) guarantee poor performance. It's either that or you limit their opportunities. Thus Ms. Wolf: "The beauty myth is always actually prescribing behavior and not appearance." She was

* The study of beauty by traditional methods is especially susceptible to the problem of insufficiency. If your research topic is, say, wealth, you can very easily get a measure of someone's net worth or income and then move on to the dependent trait you want to look at. But to study beauty, first you have to determine how good-looking your subjects are, which is a resource-intensive process. Beauty being so wildly subjective (as opposed to, say, hair color, where if you crowdsourced it, you might get slight variations—*brown, brunette, chestnut*—that are essentially synonymous), you get wide swings in opinion that can only be absorbed by sampling a large, diverse research set. As we've seen with WEIRDness earlier, that has not been a strength of past academic research.

speaking primarily in a sexual context, but here, we see how it plays out, with mathematical equivalence, in the workplace.

As I've mentioned before, I have a young daughter, and in our rare downtime, Reshma and I will speculate about her and her life and where it might lead. All parents do this—give them a quiet moment and it's inevitable, just like two drunks in a bar will always argue. Every family must have their own particular flights of fancy, but ours go more or less like most, I imagine. My wife or I will start, it doesn't really matter who: Our little girl's going to be so smart. Oh yes, we'll teach her everything we can. She'll be so gentle, so good-hearted. These things are very important to a good life, we agree. And of course, look at that skin, like chai, those eyes, she'll be so pretty. I mean, wow. Yeah, we'll have to put locks on the doors when she's a teenager. And there the conversation takes a little turn. But not too pretty, right? Yeah, we wouldn't want that. We both sit back, and the conversation moves on to something else. This is what it comes down to: I can't imagine anyone wishing limits on a son.

Unfortunately, it's a problem the Internet is surely making worse: for *The Beauty Myth,* social media signals Judgment Day. Your picture is attached to practically everything, certainly every résumé, every application, every byline. If people care about what you are doing, they will find out what you look like. Not because they should, but because they can—Facebook and LinkedIn have essentially extended OkCupid's Love Is Blind problem to everything. Even just ten years ago, it was almost impossible to tie the average person's name to her photograph; now you just Google the words—everyone does—and up pops a thumbnail from a social network. We've all had to pick through snapshots for that "best" one. Choose wisely, friends, because it defines you in a way it never has before. There's a momentum to the trend that might not be obvious to people who work outside the industry. The new design standard of the last two or three years, more open and more photocentric—what I think of as "Pinteresty"—is making not just pictures, but *beauty specifically* more important. OkCupid recently made a change for some photo displays, going from the size of the smaller black box to that of the gray, below:

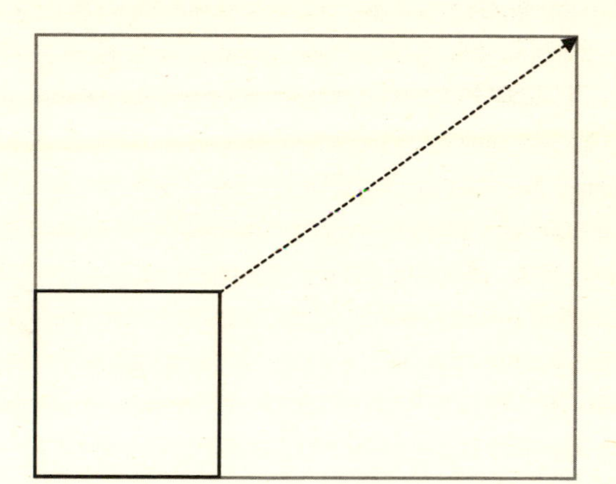

The designers just wanted the page to look more modern. What they didn't anticipate (and later had to mitigate) was the following: all those extra pixels allowed the pretty faces to outshine the others all the more. The rich got richer. It was the web-design equivalent of American domestic policy.

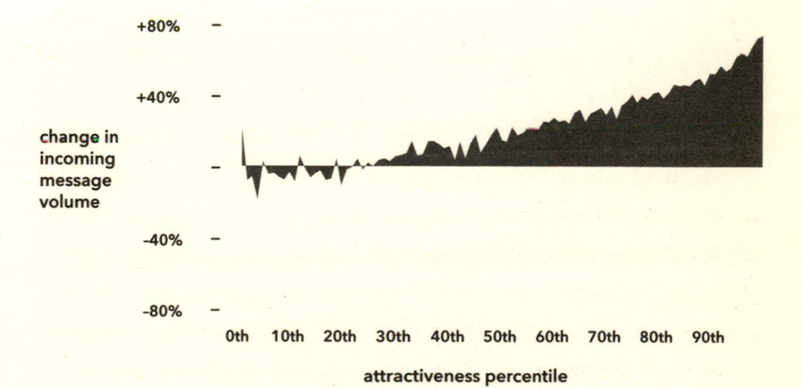

Given this pressure it's no wonder that body-image blogs are so prevalent. And that posts tagged like #thinspiration #thinspo #loseweight

#keeplosing #proana #thighgap became so common that both Tumblr and Pinterest (independent of each other) had to alter their Terms of Service to ban this kind of content. If you're wondering what the last two hashtags are, #proana is short for "pro anorexia"—people *in favor of* starvation as a weight-loss technique. Meanwhile, #thighgap refers to having thighs so thin that they do not touch when you stand with your feet and knees together. It's a trait fetishized by teenage girls. Quite apart from the questionable desirability, it's biologically impossible for most of them. The full depravity of the phenomenon can't hit you until you search for these tags yourself and are confronted with an unending page of broken bodies tilting at the camera—not only are the "inspiring" women deathly thin, they are also frequently in lingerie, bikinis, underwear. The blogs, created by women, are truly the epitome of the male gaze—and I say this as a person reflexively skeptical of the language of the academic left.

Tumblr and Pinterest banning the content didn't solve anything, of course, least of all their users' body-image issues, so the sites are now taking another approach. Because these blogs are tagged, they are able to intervene algorithmically—search for thighgap on Tumblr and the screen goes blank, an overlay appearing:

> "if you or someone you know is dealing with an eating disorder . . ."

A link to help and resources follows. It is a small measure, but before the behavior was digitized, there was practically no way to get directly at this problem, at least not until visible damage had already occurred. There was only rumor—an ear at the bathroom door, perhaps a parent's sad suspicion. Data is about how we're really feeling—feeling about one another, yes, but also about ourselves. If it finds divides in our culture, our politics, our habits, our tribes, it finds divides within us, too. And that's a hopeful thought, because for anything to be made whole, the first step is to know what's missing.

It's What's Inside That Counts

There used to be two ways to figure out what a person really thinks. One, you caught her in an unguarded moment. You snooped around, you provoked, you constructed some pretext in a laboratory, you did whatever you could do to get your subject to forget she was being watched. Research like that was probably a lot of fun—a lab coat, a hidden camera . . . who knows, a fake mustache—but on a large scale, it was impossible. So for data en masse, you had only option two: ask a question and hope for an honest answer. That's been the popular standard since Gallup formed the American Institute of Public Opinion in 1935.

Unfortunately, surveys have historically been unable to uncover true attitudes on topics such as race, sexual behavior, drug use, and even bodily functions, because respondents edit their answers. Observed behavioral data is very useful, as we've already seen. But there are some things—thoughts, beliefs—that don't entail an explicit action. And often the ugliest, most divisive, attitudes remain behind a veil of ego and cultural norms that is almost impossible to draw back, at least through direct questioning. It's a social scientist's curse—what you most want to get at is exactly what your subjects are most eager to hide. This tendency is called *social desirability bias*, and it's well documented: the world over, respondents answer questions in ways that make them look good. The most famous case was the so-called Bradley effect: in 1982, California voters told exit pollsters they had elected a black governor, Tom Bradley, by a significant margin, but in the privacy of the ballot box they had actually given his white opponent a narrow victory. Throughout the '80s and '90s, black candidates often received more support in polls than in actual elections. Problems beyond racism, like depression and addiction, are similarly difficult to diagnose at a societal level because people can't be honest about them. Even on OkCupid's match questions—which are by and large unseen by anyone but the answerer—the users are just unwilling to own up to certain attitudes, even ones they in fact act upon elsewhere on the site. The mere act of asking elicits self-censoring. Almost every site that registers opinions or collects descriptive data has the same problem. But there is one place that doesn't need to ask for anything, and so the data is set free: With search, there is no ask. You just tell.

Google's only prompt is that famously open page, with its lone entry form—that slim rectangle of emptiness, cursor parked and ready, just waiting for your thoughts. The company's business is to help people find stuff in the vast thicket of the Internet, and it's done that spectacularly. But almost as an afterthought to its world-beating success, as users enter each new desire into the database, Google has become a repository for humanity's collective id. It hears our confessions, our concerns, our secrets. It's doctor, priest, psychiatrist, confidante, and above all, Google doesn't have to ask us for a thing, because the question is always implied in the blank space of the interface: *Hey, what's on your mind?* Ahab and his whale, Arthur and his grail. What a person searches for often gives you the person himself. The trick till now has been, How can we see the search?

Since 2008, Google has provided that insight with its Google Trends tool. It allows anyone to query their aggregated search database, and with the right phrasing and a little cross-tabulation, you can use it to extract an excellent sample of the private mind, of the internal workings that have until now remained off-limits to research since research began. Since the service launched, scientists have used Google Trends to predict the stock market, uncover drivers of economic productivity (richer countries are more concerned with the future than the past), and most famously, track epidemics of flu and dengue fever in real time—and thereby stanch them as quickly. When people are getting sick, they search for symptoms and remedies. Google Flu senses what's afoot and alerts the CDC.

The site also records other kinds of virulence. Because there is no asking, and unlike on social sites, no other person on the other end of the line, people unleash their vilest impulses into Google. "Nigger," for example, is a common search term—included in 7 million searches a year. In the United States, the search volume is highest where you might expect—West Virginia—but it's steady throughout the country. Brooklyn has few things in common with the town I grew up in, Little Rock, but this is one—"nigger" is as common in New York City as it is in central

Arkansas, and as common in Chicago as it is in Fresno.* Judging by search volume, the word is literally more American than "apple pie"—by 30 percent. And, tellingly, it appears much more often in Google than it does in a more public venue for the psyche, Twitter. Using "nigga" as a control, since it's similar in meaning but lacks the baggage, "nigger" appears about 30 times more often in search than in social media.

Unlike the acute cycles of disease, racism runs a slow, grinding course—working at the generational, not the metabolic level—and it's one of the few places where we can begin to see data's broad longitudinal possibilities. Further, tying the ebb and flow in searches to real-world events allows us to unlock some of the emotional shading behind the data. For example, if you plot searches for the word "nigger" over the 2008 campaign cycle, you can watch the country come to grips with the prospect of a black president.

change in Google searches for "nigger," December 2007-February 2009

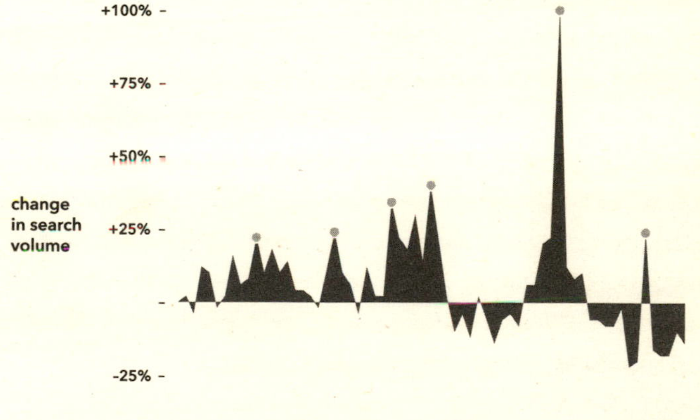

* Google Trends expresses a search's popularity with a simple index number proportional to the number of searches for the word or phrase. The indices for this epithet are within 10 percent of each other for the listed metro areas. "Nigga" is not included, since most of its related searches are for rap lyrics (the exact search query for my data throughout this chapter was: "nigger –nigga –song"). The top related searches for "nigger" are, by far, "jokes" and "nigger jokes." For my racial search analysis, I'm relying on a method originated by Seth Stephens-Davidowitz, a data scientist and economist at Google. Reporting from his inside view of the data, he writes: "A huge proportion of the searches [for "nigger"] were for jokes about African Americans." He uses public and anonymous data for his research.

Working through the six gray-dotted peaks, from left to right you see: Super Tuesday on February 5, followed by the bitterly contested Pennsylvania primary on April 22. On June 6, searches hit a new high. Hillary suspended her campaign, and Obama won the nomination. On July 15, complicating the data (and indeed the moral discussion), Nas released an album whose unofficial title was *Nigger*, and it went to number one. But even in the wake of that confounding event, overall search volume plummeted as the fact of Obama's ascendancy settled in. Racial and even political tension dissipated while the nominees, neither yet official, positioned themselves for the fall. In fact, the volume of racially charged searches reached its lowest point over the whole campaign the week of the Republican National Convention in early September.*

Having hit a minimum there, however, animus built back quickly to the norm, then exploded on election night itself, when searches for "nigger" hit a level never since equaled. The next day, when America woke up to the confirmed reality of a black president, roughly 1 in 100 searches for "Obama" also included the epithet or "KKK" in the query string. But almost immediately afterward the volume of racially charged searches dropped sharply, and except for one last gasp of anger at the inauguration, that lower level (25 percent below the pre-Obama status quo) has held. You hear a lot about our "national conversation" on race; when you look at the data, you see it's really more a series of national convulsions. But you also see that for all the failed promise of his famous byword, Obama did change the course of our nation's favorite epithet:

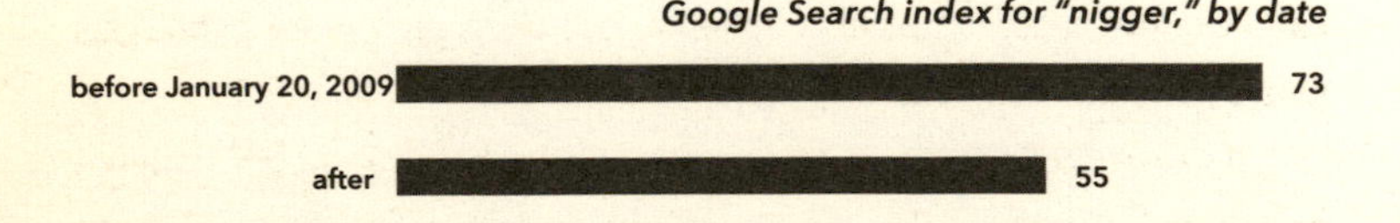

* This wasn't just people going on vacation: neutral terms like "pasta," "pizza," "family," and "truck" hold steady throughout the year.

There have been, in fact, only three true jumps in "nigger" searches during the Obama presidency. The first was driven by the kind of what-the-fact that Tea Party politicians seem to specialize in: volume spiked in October 2011, the week the world discovered that Texas governor Rick Perry has a "Niggerhead Lake" on his property. The remaining two peaks, both comparable to Obama's election night in height and suddenness, were the bookends to a single story. The first hit the servers in late March 2012, and the other the last week of June the following year. They coincide with, first, Trayvon Martin's parents bringing their son's death to national attention, and, second, when the prosecution made its case against George Zimmerman—perhaps the two times since Obama's first campaign that whiteness felt most attacked. There was no comparable spike during the defense phase of the proceedings, nor at the verdict. And, like they did in the aftermath of the 2008 election, searches hit a new low right after the acquittal, again showing the cycle of clench and catharsis that passes for race relations in the United States.

When you're out hunting for racially charged words, "nigger" is the obvious place to start, but very quickly you find there isn't much else of significance out there; it's really the alpha and omega of hate speech. Other awful terms like "spic" and "chink" are so seldom used that there's comparatively little data to analyze. It's not the epithets themselves that are the most meaningful, anyhow—it's the mind-set behind them, a truth you can see in the way the freight of the word "nigger" changes with the identity of the speaker. If it were Toby Keith and not Nas releasing that album in 2008, you'd have a *much* different story on your hands. To that end, Google's autocomplete function is useful; it gives whole thoughts rather than just a context-free word.

If you're not familiar with autocomplete, when you begin typing a phrase, for example "Who is the . . ." Google offers to finish your thought with the text from other popular searches. Type in "Who is the . . ." and it suggests ". . . richest man in the world." Tinker with it a bit, and it'll give you a peek at humanity wondering how the other half lives.

Why do women . . .
. . . cheat?
. . . have periods?
. . . wear high heels?
Why do men . . .
. . . pull away?
. . . fall in love?
. . . lie?

And when you start fishing for stereotypes, it's like playing the game Taboo, but without any taboos. Why do black people . . . like fried chicken? Why do Muslims . . . hate America? Why do Asians . . . look alike? Autocomplete gives you this kind of stuff—those are verbatim examples. In fact, one such result, "Why Do White People Have Thin Lips?," is the title of a recent research paper that explores the dual purpose the feature serves: it reveals trends, of course, but because of Google's ubiquity it has the power to set them as well. The paper suggests that autocomplete will eventually perpetuate the stereotypes it should only reflect, and it's easy to see how: a user types an unrelated question, only to have other people's prejudices jump in the way. For example, "Why do gay . . . couples look alike?" was not a stereotype I was aware of until just now. It's the site acting not as Big Brother but as Older Brother, giving you mental cigarettes.

When you turn the autocomplete queries inward, you get still another view of humanity. It's like standing alongside someone in front of his bathroom mirror. Go to your search bar with:

"Why is my a . . ." then
"Why is my b . . ." and so on

and Google will complete your prompts with an alphabet of troubles, including this brilliant run:

why is my <u>s</u>tool green
why is my <u>t</u>ongue white
why is my <u>u</u>rine cloudy
why is my <u>v</u>agina itchy

All of which ailments, I have to point out, are probably the result of sitting at a computer for too damn long.

So in all these ragged ways, our hidden thoughts are becoming part of the world. With a little creative typing, a few workarounds, and some math, we are giving humanity's inner monologue a wider audience. We bring out the hurtful as well as the ridiculous parts of ourselves, and for those hurtful impulses, search data provides much-needed exposure. It is no longer publicly acceptable to say racist things, but we can now know they're still being spoken even when social desirability bias might tell us otherwise. Moreover, though our power to detect latent, hidden attitudes is new, our power to exploit them is not, which is why this data is all the more important. I'll let Republican strategist Lee Atwater explain; below he's discussing his party's so-called Southern Strategy in an interview with political scientist Alexander P. Lamis. He said this in 1981, as a member of the Reagan administration:

> You start out in 1954 by saying, "Nigger, nigger, nigger." By 1968, you can't say "nigger"—that hurts you. Backfires. So you say stuff like forced busing, states' rights and all that stuff. You're getting so abstract now [that] you're talking about cutting taxes, and all these things you're talking about are totally economic things and a by-product of them is [that] blacks get hurt worse than whites.

Atwater thought he was speaking off the record ("Now, y'all aren't quoting me on this?"). Search data means we don't have to wait for such accidents to examine the disconnect between the public and private conversation on a topic like race. It shows we're heading toward a better world. It also shows we have far to go.

Let's pick up where we left Obama, on Inauguration Day, 2009. There was a lot of hopeful talk then that the United States had become a "post-racial" society, and it wasn't necessarily a far-fetched idea. At its core, the "post-racial" story was an attempt to extrapolate the success of Obama's campaign to other corners of American life, and to say that his victory proved that "race wasn't a factor" in our lives, not anymore.

Despite that hopeful possibility, Seth Stephens-Davidowitz at Google concluded that Obama's race probably cost him 3 to 5 percentage points of the popular vote in 2008—and the loss wasn't from Republicans but from people who otherwise would've voted for a white Democrat like John Kerry. At the high end of the range, that 5 percent swing would've altered well over half the elections since World War II, and it's a result we could never have detected without search data. The researcher's brainstorm was to go back *before* Obama entered the national political picture, to 2004–2007, and mine Google Trends for preexisting racial attitudes. (That keeps dislike of Obama himself from clouding the picture.) Using that data to get a state-by-state "racial animus index," he could then compare that index against Obama's eventual vote totals and against the expected outcome for a generic (i.e., white) Democratic candidate (for which of course there is ample previous data). Reliably, the higher the animus index, the worse Obama performed. Here's an example of the method in the words of the man who did the work:

> Consider two media markets, Denver and Wheeling (which is a market evenly split between Ohio and West Virginia). Mr. Kerry received roughly 50 percent of the votes in both markets. Based on the large gains for Democrats in 2008, Mr. Obama should have received about 57 percent of votes in both Denver and Wheeling. Denver and Wheeling, though, exhibit different racial attitudes. Denver had the fourth lowest racially charged

> search rate in the country. Mr. Obama won 57 percent of the vote there, just as predicted. Wheeling had the seventh highest racially charged search rate in the country. Mr. Obama won less than 48 percent of the Wheeling vote.

Historically, a presidential candidate can expect a modest boost, about 2 percentage points, in the popular vote in his home state. Because of racial animus, John McCain in 2008 had better than home-state advantage throughout the entire country. If you're looking for evidence of whiteness as a leg-up in American life, this is it. McCain was the nation's favorite son for no other reason than he was pitted against a black man.

In my opinion, Muhammad Ali is one of the bravest Americans. In 1967, as heavyweight champion, he refused to serve in Vietnam and was not only stripped of his title but banned from the sport for three and a half years. He lost the prime of his career, and received a five-year prison sentence (that took the Supreme Court to overturn), because of what he believed in. It's a stand unimaginable from today's political leaders, let alone our athletes and celebrities. From Kanye to Glenn Beck to Rachel Maddow to Sarah Palin, you get plenty of anger, but little sacrifice. We can each have our own take on Ali's stance against Vietnam—and as the son of a veteran, Huế '69, I know at least one person who disagrees with mine—but data like this can help anyone understand *why* he took it. As Ali said at the time, "No Viet-Cong ever called me nigger," and he was probably right. But imagine, had Google existed then, what would've been going into American search bars. And imagine the home-state disadvantage of a black man in those days.

It remains to be seen where attitudes will go next. For all the above, Obama *did* win, and as depressing as some of this stuff is, there's a lot to be encouraged about—for one thing, there was no evidence that bias hurt the president again in 2012, though he was a known quantity by then, perhaps less "a black man" than "Barack Obama." One thing that gets lost in all the aggregation throughout this book is that on an individual level, the personal effects of these broad social forces are often very subtle. To speak to the data you've seen in a previous chapter, OkCupid's many black users have a fine experience on the site—each one of them gets dates and rejection like anyone else. They just get, collectively, more of the latter. When you go person-by-person, any individual's experience is too small and too varied to *conclusively* say anything "racial" has happened. It could be your skin, or it could be just you. On the other side of it, it's laughable to think of one red-faced guy searching for "nigger jokes" because Barack Obama got elected. But it's a lot less funny when you can see that he's one of thousands and thousands making the same search. And it's less funny still when you see the large effect these private attitudes can still have, even in public life. Thus the story of just one of us versus the story of us all. That's why data like this is necessary—it ends arguments that anecdotes could never win. It provides facts that need facing.

I know some people who only read good books—and by that I mean things that come recommended: by friends, teachers, reviewers, Amazon. It makes sense; reading is slow, time is precious, why risk it? But that's not my style. I like history, and when I go to the bookstore, I just grab a bunch of random stuff from the section shelves and see what sticks. Reader, I have read some bullshit. And too many books on Napoleon. But among many serendipitous discoveries, *A People's History of the United States* is my favorite. Yes, I know now it's a classic, but that doesn't change the fact that I'd never heard of the book until I pulled it down. Google Books describes it well: it's a chronicle of "American history from the bottom up"—and where most books treat leaders and big events, *A People's History* shows us the homes, shops, farms, factories, and smaller worries of yesteryear. The thing is, as much as I love that book, and as much as it turns the schoolhouse version of American history on its head, Howard Zinn could still only tell us what he could see, the observable actions, the words spoken aloud. The hearts of women and men were beyond him. In the stress of the Cuban Missile Crisis, in the boredom of the trenches, in the liberation of the Pill—for all the moments of quiet joy and interior anguish lost to history, what if we had the data we have now? How much richer would our understanding be?

9.

Days of Rage

On New Year's Eve, bored on her couch and waiting for the ball to drop, Safiyyah Nawaz tweeted a silly joke.

> $afiyyah @safiyyahn
> this beautiful earth is now 2014 years old, amazing

She got 16,000 retweets, almost all of them in the next twenty-four hours. For reference, Katy Perry's Happy New Year wish to her 49 million followers got just over 19,000. Lady Gaga's, which also announced a long-awaited video, got 20,000. Safiyyah Nawaz is not some emerging world pop star, and this isn't the story of Twitter empowering upstarts to challenge the cultural order. If you haven't heard of Safiyyah, that's because she's a North Carolina high school student whose joke, the exact words above, made Twitter explode.

At first it was people verbally scratching their heads, wondering if she was serious, but if you watch the tweets from that night go by, each retweeter a further degree removed from Safiyyah the human being, and each more aware that his or her ridicule was part of a phenomenon—this from watching the retweet number tick up—you can actually see the digital crowd become a mob. In short order, the amused LOLs became OMGs became WTFs, and then stuff like this took over:

> Cocaine Burger @Cocaine_Burger
> @safiyyahn Kill yourself

> Rick Huijbers @HARDEBAKSTEEN
> @safiyyahn kill yourself you stupid motherfuck

It went, as *Gawker* put it in their coverage, from *dumb* to *#dumbbitch* in a matter of minutes. Given the violence of the reaction, Ms. Nawaz handled the experience pretty well for a seventeen-year-old, and later she sized up the outcry perfectly:

> $afiyyahn @safiyyahn
> young folks these days b really passionate about the tru age of the earth

Nawaz was unaware of it, but she had famous company in the cross-hairs. Just fifteen minutes before she'd tweeted her joke, comedian Natasha Leggero was in Times Square, on television with Carson Daly, bantering about the SpaghettiOs Pearl Harbor Day PR campaign. The brand had come under fire for encouraging citizens to remember the fallen via purchase of canned spaghetti—yes, this is what the world has come to—and she said, "It sucks that the only survivors of Pearl Harbor are being mocked by the only food they can still chew."

Host and guest laughed and moved on to other things, unaware that Natasha, too, had inadvertently brushed against the highly sensitive On switch of the Internet-rage machine. It sputtered into righteous action; Ms. Leggero later posted on Tumblr several choice examples of the tweets she got. Stuff like:

> Mike Oswald @SDPStudio
> @natashaleggero What a vile whore you are.

> Mark Tichenor @hotrod607
> @natashaleggero Fuck You, you disrespectful cunt

And my personal favorite, which, should the Internet ever die, will be its epitaph:

Chris McAllister @macdawg22

@natashaleggero your a stupid ignorant whore.

I was paying special attention to these two episodes because something similar had just happened to a coworker of mine. On December 20, Justine Sacco, who was director of communications at OkCupid's parent company, IAC, was at Heathrow, waiting for a connecting flight to Johannesburg. She boarded the plane, sat down in her seat, and typed:

Justine Sacco @justinesacco

Going to Africa. Hope I don't get AIDS. Just kidding. I'm white!

Then she turned off her phone. Her tweet was less obviously a joke than the other two examples and at best—at best—it was a clumsy dig at white privilege. But what started with justified head-shaking at her cluelessness quickly became a carnival of intense personal hatred. She got the usual threats and insults, but the attack aimed for more than her Twitter persona. Pictures of her family were circulated online, along with their whereabouts. Men called her nephews, threatening to rape them. People gathered at the Johannesburg airport to await her plane. Her inability to respond while aloft added an extra jolt of enthusiasm to the takedown. About midway through her flight #HasJustineLandedYet was coined and became a top trending topic on Twitter. Google searches for her name began to automatically return her flight number and its arrival time because that's what people were searching for—search algorithms had again held up a mirror. For the eleven hours Justine hung in the air, the Internet waited dry-mouthed and bloodthirsty for the moment she would reconnect to find her life in ruins.

Ron Geraci @RonGeraci

It's like 2 million people are waiting for her with the lights off to see her expression as the earth explodes.

I'm Gary @noyokono

#HasJustineLandedYet People haven't eagerly anticipated a plane landing this much since Amelia Earhart.

V. Hussein Savage @Kennymack1971

Aw hell. . . . lemme finish this work grab a 6 pack and some BBQ wings. It's about to be on. . .
#HasJustineLandedYet

Their quarry here was someone with a few hundred followers and no public profile. I didn't know Justine all that well, but I had enjoyed working with her, and watching the obvious excitement people got from the pain and fear they were about to cause sickened me.

Like a fool, I went to Facebook to vent. My post wasn't up ten minutes before an acquaintance (and future former Facebook friend), who at that point I hadn't spoken to in fifteen years, commented "her father is a billionaire" and implied that that somehow justified her personal destruction.* But of course her father isn't a billionaire—that was just another rumor that had attached itself to the story. It was like running into a mob at a stoning, trying to drag people away, finding someone you know—whew, finally, a guy you can reason with—only to have him yell, wide-eyed, "Dude, check out all these rocks!"

The stoning metaphor comes up again and again when you read the commentary on episodes like these. It's no coincidence that it's the death penalty of choice for the ancient religions: there is no single executioner; the community carries out the punishment. No one can say who struck the fatal blow, because everyone did together.† For a burgeoning tribe, fighting to preserve itself and its god in a hostile world, what better prescription could there be? There is strength in collective guilt, and guilt is

---

* If Facebook ever gets tired of that minimalist *f* and wants a new logo, I suggest, on a blue background: two white people arguing about what another white person said about Africa.

† It would be interesting to see if residents of countries where stoning is still used as a real-world punishment take as much joy in the digital version.

diffused in the sharing. Extirpate the Other and make yourselves whole again.

In Justine's case, people on three continents had assembled to destroy her. Pulling self-descriptions from just a handful of their Twitter bios you find it takes all types: Lobbyist. Communist. Hater. Aspie. Leader. Nature Enthusiast. Blogger. Gator. Dad. Writer. Imperfect Christian. Professional Shade Detector. Pop Culture Virtuoso. Daughter of the Sea, Sister to the Wind. These people had nothing in common but a target and a hashtag at hand, and they got the blood they came for. Justine lost her job. BuzzFeed put her face up on their front page with a big "LOL" over it.

The reach of social media makes the force of these gatherings immense. Within twenty-four hours of her tweet, Safiyyah had been called down in front of 7.4 million people. And 62 million saw #Has JustineLandedYet that first day.

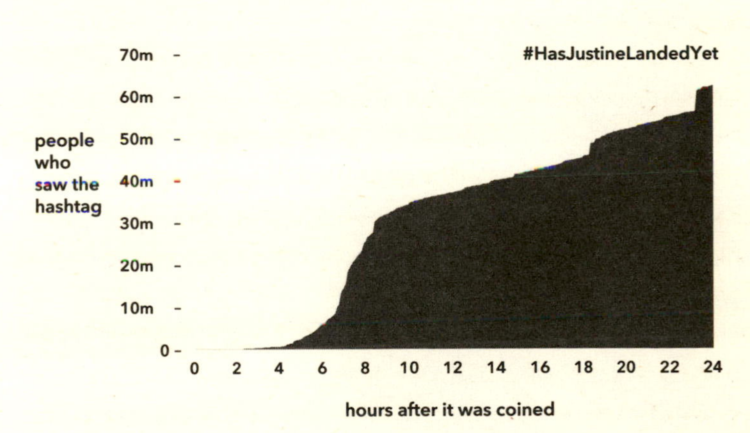

Not everyone under the curve read the tweets or cared, but many did, and all were in some way a witness.

Sir Qwap Qwap @BeardedHistoria
Literally every one of the first 20 tweets on my home feed has #HasJustineLandedYet. I must have missed something, Tweet-fiends.

It's worth pointing out that this fantastic volume should be an embarrassment to social media—evidence not just of its power but of how hollow that power can be. In Justine's case, AIDS, racism, and the stubborn, shameful poverty of postcolonial Africa are all enormous problems that tweeting does absolutely nothing to solve.

We may think of human sacrifice as something from a savage past, and the physical act might now only exist in films about temples and doom, but the instinct remains within us, seemingly burned by deep time into the reaches of the animal mind. When food is scarce, lions kill their cubs. Fish eat their own eggs. In multiple human pregnancies a womb will sometimes absorb a fetus to preserve the others. To destroy the one for the many is possibly a practice as old as life itself. Now that this ritual is carried out in bits (and thankfully with no actual blood on anyone's hands, though you get the idea, reading some of these tweets, that people view this as a bug rather than a feature), it's become a topic we can rigorously study for the first time. Social scientists have devoted considerable energy to the question of why and how negative ideas spread, and the Internet has given them both limitless source material and a powerful tracking mechanism. Marine biologists tag sharks in the wild to understand their movements and to limit their threat to humans.* Here it's the words that have teeth. My three cases above aren't precisely rumors or gossip, but mob outrage follows many of the same pathways, both neurological and person-to-person, and the science of rumors can help us understand what has happened to people like Natasha, Safiyyah, and Justine—and why.

Rumors are mentioned in our earliest texts. The archaic pantheons—Norse, Egyptian, Greek—all have a god dedicated to the dark art of gossip. The book of Proverbs treats the topic thoroughly; one verse from many cautions that "a man who lacks judgment derides his neighbor, but a man of understanding holds his tongue." "Judge not lest you be judged" is one of

* In Australia these tags are outfitted with transponders that notify local beachgoers when a shark is nearby. The tags communicate to us . . . via Twitter.

the most famous phrases in the whole Bible. Several sources maintain that the Romans enshrined a goddess named "Rumor"—a winged demon with a hundred eyes and a hundred mouths who spoke only the most hurtful side of the truth. Appropriately enough, I can't seem to confirm this.

Evolutionary biologists believe that gossip and rumors arose from our ancestors' need to understand their surroundings through speech. The theory is, when ancient man had to figure out if *x* was true, language gave him a way to investigate. So he talked about it. And, true or false, word spread. Rumors—essentially group speculation over the truth of an idea—became a way to build bonds and social capital. Stories create status for those who share them, especially when they concern important individuals, because information about powerful people is a form of power itself.

But the advent of social media has changed the calculus in a couple ways. First, it gives us metrics—follower counts, retweet counts, favorites counts—to judge our status. Be the first to spread the news, get more retweets. Say something especially cutting, and your followers applaud your wit. The social capital you build by sharing information is now explicit; in fact, it's in little numbers that increment before your very eyes. Writing in the *Boston Globe,* Jesse Singal was discussing the motivations of traditional person-to-person gossip but might've easily been talking about Twitter when he said, "To the extent people do have an agenda in spreading rumors it's directed more at the people they're spreading them to, rather than at the subject of the rumor." The Internet gives people a wider audience than ever before.

The second change is that the Internet has also made everyone a public figure. High-status individuals were once chieftains, and then celebrities and presidents, but here, the leveling scythe of technology shows its obverse edge. If anyone can become an overnight celebrity, anyone can become an overnight leper. One of my least favorite Internet-evangelist talking points is about technology "empowering" people—inevitably the most empowered of all is the speaker and his investors. But here we find some truth in the cliché—social media empowers you to the extent that it makes you worth tearing down. At the same time, it gives everyone else the tools to do it. Demon Rumor now has a million mouths.

So much of what makes the Internet useful for communication—asynchrony, anonymity, escapism, a lack of central authority—also makes it frightening. People can act however they want (and say whatever they want) without consequences, a phenomenon first studied by John Suler, a professor of psychology at Rider University. His name for it is the "online disinhibition effect." The webcomic *Penny Arcade* puts it a little better:

> Greater Internet Fuckwad Theory
> normal person + anonymity + audience = total fuckwad

But it's not the vitriol, nor even the anonymity, that's unique here. The Internet hasn't been quite the revolution in trollery you'd think. The old CB radio channels that truckers used were notoriously filled with racist diatribes and masturbation fantasy.* Before caller ID took away that necessary additive, anonymity, the Jerky Boys were churning out fuckwaddedness for decades. People still flame one another on ham radio—as if being a ham radio operator in 2014 isn't burn enough. No, the unique thing that the Internet brings to our long history of negativity is that we can finally constructively respond to it. In some way, Tumblr's thighgap intervention discussed in chapter 7 is just a special case of what's now broadly possible. We can pinpoint the speaker, the words, the moment, even the latitude and longitude of human communication. As I pointed out earlier, by 2015, Twitter users will have exchanged more words than have ever been printed. The question is how to harness the chatter.

The government has the greatest vested interest in tracking negativity. Mathematical models already exist to predict the outcome of armed

* And, as they do online, the users even had "handles."

conflict—how long it will last, who will win, and how many people will die—and the models of late have learned to accommodate guerrilla warfare, since that's the shape of today's war. But armed insurgency is often preceded by *unarmed* unrest—which itself is often propagated, even coordinated, through social media.* Those nascent movements, being digitized, have attracted the attention of researchers.

Using Western movements as his test subjects, MIT's Peter Gloor has developed software to track the ebb and flow of sentiment in a network of protestors. He calls it Condor, because that's what projects like this always seem to be called: Condor, spirit-bird of government grants. In any event, the software first establishes a group's central personalities by looking at its social graph—much like we portrayed a marriage as edges and nodes before, the software lays out the network, then algorithmically determines its most important dots. Next, it looks at what those dots are saying. Condor has found that while the foci of a movement are positive in their word choice, the movement is vibrant. But negative words like "hate," "not," "lame," and "never" signal decline, and when, as *The Economist* put it, "complaints about idiots in one's own movement or such infelicities as the theft of beer by a fellow demonstrator" begin to appear, the movement is all but over. Oh, Occupy!

As for deciphering the *aims* of unrest, which is where this technology can move beyond mere spying and into doing some good, similar kinds of textual analysis have been used to determine, for example, which Egyptian towns will be most upset by border incidents with Israel, and to pinpoint water insecurity in a drought-stricken countryside.

Any software that follows the thread of a thought through a network must track not only the idea but the "susceptibility" of people exposed to it. It must see what takes hold, what gets repeated, and who moves it along. Relaying someone else's opinion isn't unique to the Internet any more than negativity is: television and radio made "talking points" into a phrase

* The Arab Spring, for example, was Twitter's debut as a tool of global importance, and the service has also facilitated protests in Guatemala, Moldova, Russia, and Ukraine.

long before AOL came along, let alone Twitter. Rush Limbaugh's staunchest fans call *themselves* "Dittoheads"—but nothing makes parroting an idea more simple, or more trackable, than the Like, the Ping, the Reblog, or the Retweet button. Remember: 27.5 percent of Twitter's 500 million tweets a day are retweets, people just passing along someone else's thought.

Facebook's data team investigated their version of the phenomenon, tracing the evolution of a single status update from the health-care debates in 2009 through the network:

> No one should die because they cannot afford health care, and no one should go broke because they get sick. If you agree, post this as your status for the rest of the day.

This was reposted, verbatim, more than 470,000 times and also spawned 121,605 different variants, which themselves received about 800,000 more posts. Someone who didn't quite feel that the update spoke for him would change it slightly, and versions spread outward into different social circles. When you put each version against the political bias of the people posting it (–2.0 is maximally liberal, +2.0 conservative), not only do you get an interesting look at the American political spectrum—extremes of right and left, plus a center that has opted-out of the discussion—but you also see how political belief translates into words. People at the top and bottom of this list use the same framework to speak at cross-purposes:

No one should . . .	political bias of the person posting	
. . . die because they cannot afford health care . . .	-0.87	more liberal
. . . be frozen in carbonite because they couldn't pay Jabba the Hutt . . .	-0.37	
. . . die because of zombies if they cannot afford a shotgun . . .	-0.30	
. . . have to worry about dying tomorrow, but cancer patients do . . .	-0.02	
. . . be without a beer because they cannot afford one . . .	+0.22	

. . . die because the government is involved with health care . . .	**+0.88**	
. . . die because Obamacare rations their health care . . .	**+0.96**	
. . . go broke because government taxes and spends . . .	**+0.97**	**more conservative**

In 1950, at the dawn of the age of television, the American Political Science Association actually called for more polarization in national politics—the parties had grown too close together, the electorate didn't have clear choices. The APSA got their wish, and in the old genie-style, too, with plenty to regret about its granting. Now, sixty years later, we're more divided than ever, and you can track this, too, through the words. The repetition of partisan speech both in Congress itself and in print (as tracked through Google Books) correlates with political gridlock, which is at an all-time high. That we're divided might be the only thing we can, in fact, agree on.

This paradox was driven home to me when I turned to Facebook in the aftermath of Justine's tweet. In my post was a link to an article from breitbart.com—the namesake site of Tea Party instigator Andrew Breitbart. A lot about the article was regrettable, but the author was one of the only people pointing out how out-of-proportion the reaction was. I'd always imagined uncritical outrage as a vice of the political right—I'd hear about the ridiculous "War on Christmas" or the belief that Obama was "taking people's guns away" and think, What fools these people are to believe this stuff! Why talk about things in such extreme terms? Why look at something only in the worst possible light? But it took this incident on Twitter to make me see that people on the "left" could be just as self-righteously uninformed as anyone else. It was eye-opening, and shame on me for having them closed in the first place.

So theories aside—and the science is so new that no doubt Condor will look like Zork in a few years—this, to me, is why the data generated from outrage could ultimately be so important. It embodies (and therefore lets us study) the contradictions inherent in us all. It shows we fight hardest against those who can least fight back. And, above all, it runs to ground our age-old desire to raise ourselves up by putting other people down. Scientists have established that the drive is as old as time, but this doesn't mean they understand it yet. As Gandhi put it, "It has always been a mystery to me how men can feel themselves honored by the humiliation of their fellow beings."

I invite you to imagine when it will be a mystery no more. That will be the real transformation—to know not just that people *are* cruel, and in what amounts, and when, but *why*. Why we search for "nigger jokes" when a black man wins; why inspiration is hollow-eyed, stripped, and, above all, #thin; why people scream at each other about the true age of the earth. And why we seem to define ourselves as much by what we hate as by what we love.

PART 3

What Makes Us Who We Are

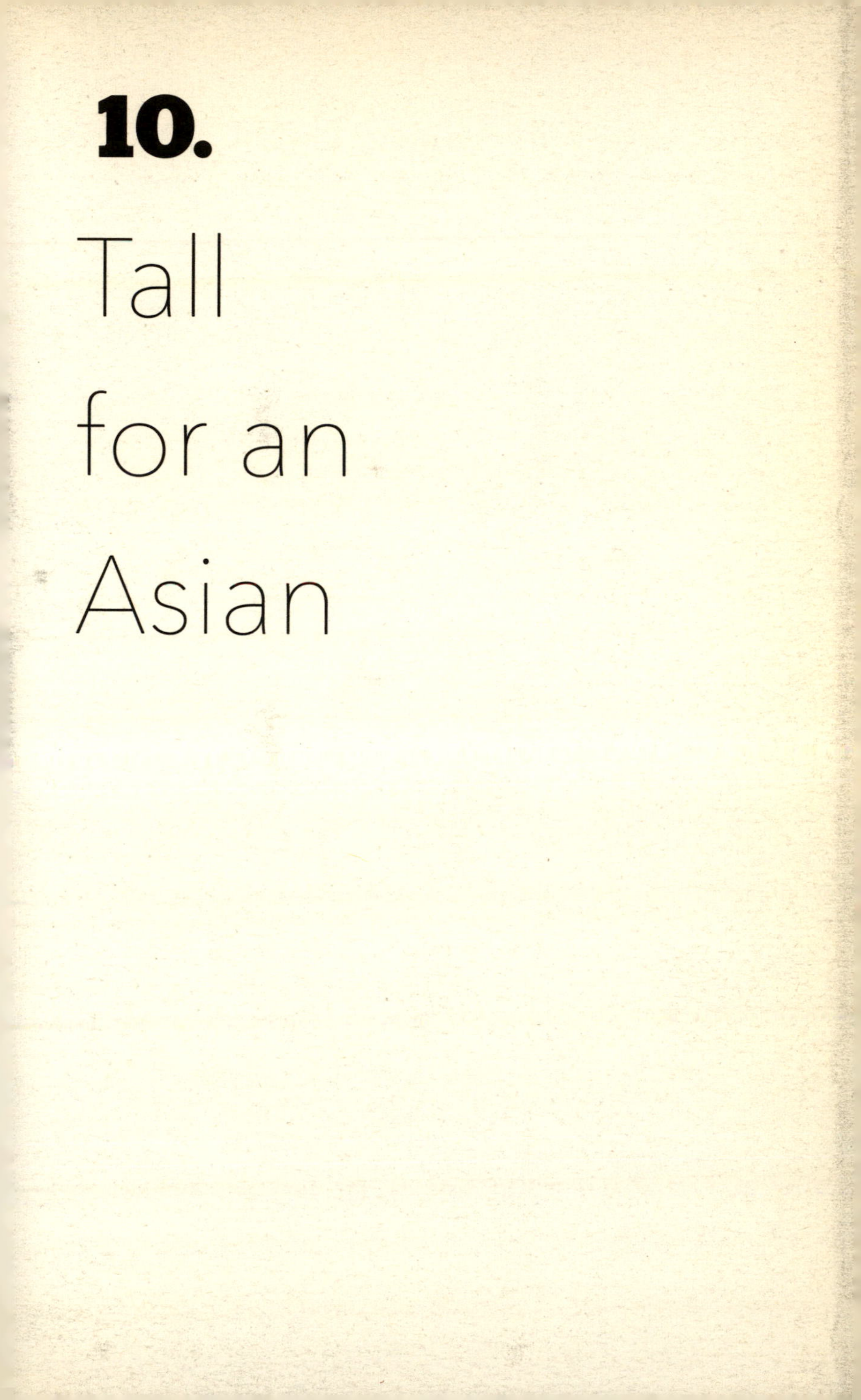

10.

Tall for an Asian

When I was applying to college, I had to write about myself. I'm sure you did, too. I can't even remember the question on the application because whatever it was actually asking was beside the point. The essay was there to get me to talk about Christian Rudder, so the Admissions people could decide if they liked what they heard. As the Common Application now puts it: "The personal essay helps us become acquainted with you as a person."

Being a sucker for melodrama even then, I wrote about how sad I would be to leave my dog behind when I went to school. We'd gotten Frosty when I was six, so he and I had grown up together. But with dog years working like they do, he'd gotten too old too fast. My family had moved around a lot, and he was that last connection to deep childhood: clubhouses, neighborhood pools, friends; I'd left them all in Houston, or Cleveland, or Louisville, but Frosty always came with me. The next move, however, I knew I'd have to make on my own.

In any event, adrift in pathos and extra-large M. C. Escher T-shirts, I completed my college application. I haven't written many self-statements since, but involved as I am in the business of understanding people I can't help but think back on my seventeen-year-old self and the essay he chose to write. Why talk about Frosty and getting older? Why not talk about baseball? Or basketball? Or tennis? Or rotisserie baseball? Or any other of my diverse interests? What was it, when the prompt was "Who are you?" that made me respond like I did? And, even more important, how were other kids answering the question?

Now, twenty years later, I find myself sitting on millions of essays—billions of words—more or less written to answer that same prompt: "Who are you?" And this body of text actually allows me to do the inverse of the college application process. Instead of matching essays one at a time against a preconceived ideal (i.e., "college material"), I can mush all the essays together and see what ideals *they reveal to me*. There are times when a data set is so robust that if you set up your analysis right, you don't need to ask it questions—it just tells you everything anyway. How do people describe themselves? What's important, what's typical, what's *atypical?* When everyone else gets a turn to put down in words who they are, what identities do they sketch?

We're going to look at broad categories here: black people, white people, Asians, females, males, and so on. A problem in studying any particular group is that you always bring your own prejudices and preconceptions along with you. What you choose to notice, remember, and transcribe is as much a matter of how you look as what's actually there. In social science, knowledge, like water, often takes the shape of its vessel. So if we want to take all the self-statements I've collected and pull from them a sense of who the writers are—what makes ethnicities and sexes and orientations unique—we'll need to develop an algorithm that takes the "us" out of it and leaves just the "them."

OkCupid's user-submitted profile essays are as close to personal self-summaries as you'll find. The prompts are open-ended:

"My self-summary . . ."
"I'm really good at . . ."
"The first things people usually notice about me are . . ."
"I spend a lot of time thinking about . . ."

And insofar as people try to put their best foot forward, they're not at all unlike college essays. I imagine many people approach them with the same sort of dread. There are no length restrictions, no guidelines but for the prompts. Altogether, people have given the site 3.2 billion words of self-description. Moreover, unlike other big hunks of text—say, what Google Books has collected—there are demographics behind every word: the age of the author, where she lives, her race, and so on. But deriving a group identity for, say, Asian women from the text isn't quite as easy as counting up who types what the most, which for the most part is how we've looked at text so far in this book. Counting words just gets us this:

1. the
2. of
3. and
4. . . .

and so on down the line—basically that top 100 from the Oxford English Corpus we saw before. Asian women, white men, and all English speakers use the same pronouns and articles and prepositions to talk about themselves. To find out what's actually *special* to a particular group, and to them alone, we have to sort the text a little differently.

I'll use white men as my walk-through example, because I understand them the best. The first step is to separate those white guys' essays from everyone else's. Then, in the two sets of self-descriptions—white-guy and not—we order all the words and phrases in the texts by how frequently they appear. We put them into two lists, from most popular to least, and that gives us something like the chart below. I've pulled out three examples and put them in their correct places in the line; the full lists have about 360,000 phrases each:

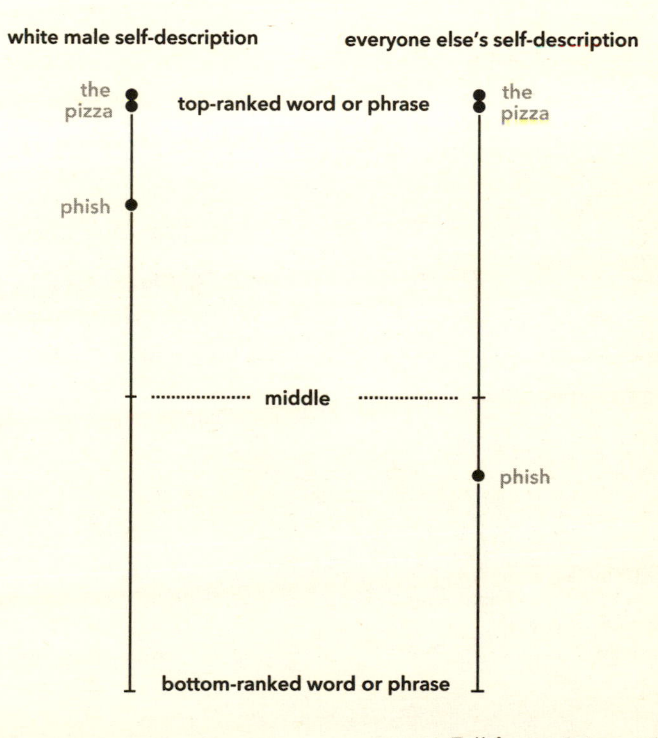

Already we're getting somewhere, but before we move on, there's something a little misleading about these plots that I want to address while the list is still simple. No, it's got nothing to do with Phish, though lord knows they've misled many. It's that "pizza" and "the" appear to be mentioned almost the same number of times. Granted, pizza is the king of foods, but "the" is the absolute most popular word in the English language. And in our data, while "the" is in its rightful place at the top, "pizza" is seemingly right there with it, at the 98th percentile. This makes it feel like something is wrong either with my data or with my method, but the rankings of the words are correct. It's just that humans use language in an odd way: we are always repeating ourselves. So a very few top-ranked words take up most of our writing. And, conversely, the frequency of a word falls off very quickly as you go even a small distance from "most popular."

This counterintuitive relationship between the popularity of a word (its *rank* in a given vocabulary) and the number of times it appears is described by something called Zipf's law, an observed statistical property of language that, like so much of the best math, lies somewhere between miracle and coincidence.* It states that in any large body of text, a word's popularity (its place in the lexicon, with 1 being the highest ranking) multiplied by the number of times it shows up, is the same for every word in the text. Or, very elegantly:

$$rank \times number = constant$$

This law holds for the Bible, the collected lyrics of '60s pop songs, and the canonical corpus of English literature (the Oxford English Corpus), and it certainly holds for profile text. To see how well it works in

---

* Another, much more famous, example is: $e^{\pi i} + 1 = 0$. Here, astoundingly, the five most important values in mathematics form a single equation. It's called the Euler Identity, by the way. He was a slacker.

practice even on a highly idiosyncratic body of writing, here's the law applied to James Joyce's *Ulysses:**

word	rank	number of times it appears	rank × number
's	10	2,826	28,260
is	20	1,435	28,700
what	30	975	29,250
has	100	289	28,900
wife	200	140	28,000
Ireland	300	90	27,000
college	1,000	26	26,000
morn	5,000	5	25,000
builder	10,000	2	20,000
Zurich	29,055	1	29,055

The steady relationship between rank and number seems to be a property of the mind as much as of language—as you can see above, it accommodates arbitrary proper names, like "Ireland" and "Zurich," and even words transcribed from dialect, like "'s."

And as further evidence of its deep connection with the human experience, Zipf's law also describes a wide variety of our social constructs: the sizes of cities, for example, and income distribution across a population. What it means for our purpose here is that because most of language is just a small body of repeated patterns, the use of a word drops off rapidly. "The" appears on nearly every profile. "Pizza" appears on about 1 in 14. "Phish," even for white guys, for whom it ranks way up at the 80th percentile, appears in less than 1 in 200 profiles. Now that we understand how rankings and usage frequency compare, the next step is to use those rankings to our advantage.

* This example is adapted from "Zipf's Law and Vocabulary," by C. Joseph Sorell, Victoria University of Wellington. Like any empirical law, Zipf's is a very good (and time-tested) descriptive framework, but as you can see there is some variance in observed outcomes. It's like knowing that a fair coin comes up heads half the time. Nonetheless, even after a thousand flips, it's very unlikely that exactly half of them will have been heads.

Below, I've put the two lists at right angles, forming a square, and I have plotted the words inside it using their popularity rankings on the two lists as coordinates. I added some arrows around "Phish" to make it clear what I mean:

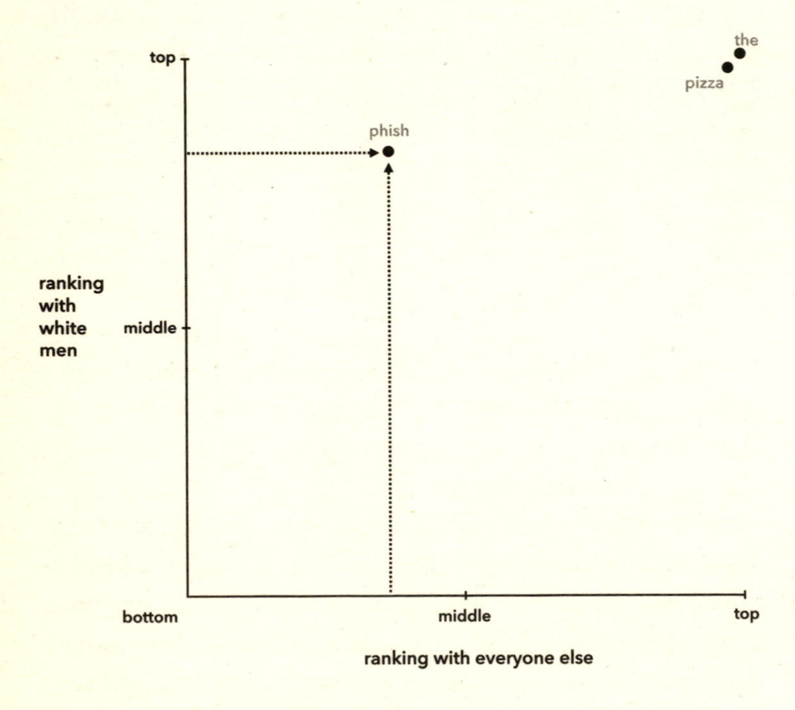

A word's position here has dual meaning. The closer to the top it appears, the more popular it is with white guys. The farther toward the right, the more popular it is with everyone else. Adding a few more words to the chart will give you a sense of how the geometry translates before I zoom out to the full corpus.

I've added a diagonal, yet again, to show parity in the data. The words near the line are important to everyone equally. And the farther up and to the right the words go, the more universally important they are. But remember, we're not looking for universals. We're looking for particulars. We want to know what is special to the people we're considering: here, white guys. For that we need to look to the upper *left*: the farther in that direction a word appears, the more

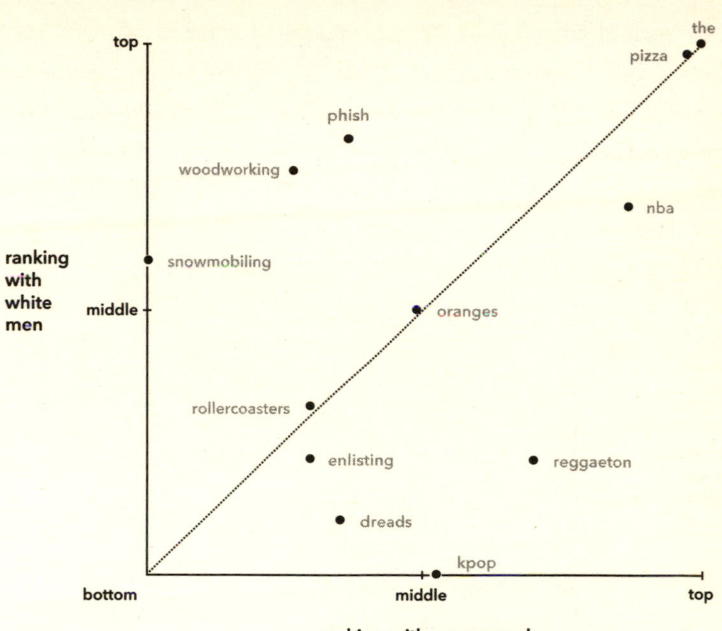

often *white men* use it, and the less often *everyone else* does. In fact, the closer a word is to that remotest reach of white maleness, the top-left vertex of the square, the more it typifies them and only them. Imagine a dot all the way in the corner: to be there, the word would have to appear on every single white male profile and at the same time *never* appear anywhere else. At least as far as words in a self-summary go, that's the platonic ideal of identity. This system, and that metric—distance from the upper-left corner—gives the data a way to speak to us, to help us understand how people are talking about themselves.

Because every data set has its quirks, researchers must often build tools from scratch, as we have here. Whenever you do this, it's good to check your method against some familiar outcomes. Imagine a shipwright with a new boat: who knows what'll happen once it's out on the open ocean—so best to check for holes close to shore. Here, if we'd found "Kpop" (Korean pop) or "dreads" in the upper left, in my supposed corner of white-manhood, it would be a strong sign that either my data or my method was garbage. But as you can see, it's working perfectly.

So, finally, here's what the whole corpus of words and phrases looks like:

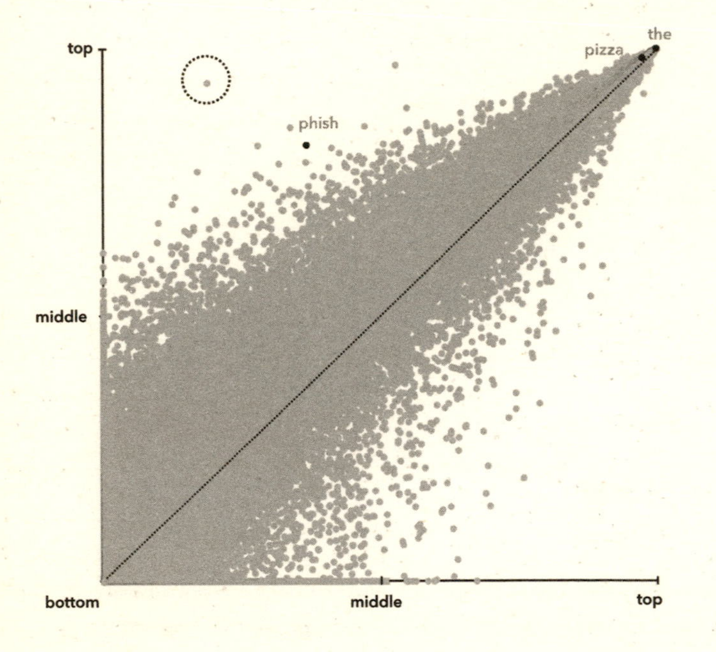

I've circled the dot closest to that upper-left corner: that's the white-male-est thing a person can write about himself: *my blue eyes*. And getting a longer list of the things that uniquely define white men is just a matter of walking out from that vertex—for example, the thirty closest dots are the thirty things that are most typical. The geometry finds the clichés for us.

I've made plots like this for everyone in my data set, not just white guys, and using this same math I've gotten lists of their unique words and phrases, too. But before I move to listing all this, I want to make one important point. Walking through each combination of sex × ethnicity × orientation gives you 2 × 4 × 3 = 24 charts like the one above, and in all of them the mass of dots has this same tapered shape from bottom left to top right. That is, the farther a phrase goes into that upper-right corner, the closer to the diagonal it gets. What that means is that *we tend to agree* *on the things that are most important.* As for the things we don't agree on, I've listed them in detail below. I'll start with the men:*

most typical words for . . .

white men	black men	Latinos	Asian men
my blue eyes	dreads	colombian	tall for an asian
blonde hair	jill scott	salsa merengue	asians
ween	haitian	cumbia	taiwanese
brown hair	soca	una	taiwan
hunting and fishing	neo soul	merengue bachata	cantonese
allman brothers	jamie foxx	mana	infernal affairs
woodworking	zane	banda	seoul
campfire	paid in full	puertorican	infernal
redneck	nigga	colombia	shanghai
dropkick murphys	luther vandross	gusta	boba
they might be giants	coldest winter	puerto rican	kbbq
brewing beer	tyler perry	tejano	kpop
robert heinlein	swagg	corridos	badminton
tom robbins	jerome	bachata merengue	kimchi
townes	dreadlocks	hector	chungking express
old crow medicine show	spike lee	espa	chou
mystery science theater	holla at me	por	viet
skis	menace to society	salsa bachata	jiro
sailboat	brotha	aventura	dash berlin
around a fire	shottas	english and spanish	ucsd
caddyshack	boomerang	musica	beijing
blond hair	nigerian	espa ol	hk
bill bryson	heartbeats	como	norwegian wood
wheelers	anthony hamilton	fiu	jiro dreams of sushi
pogues	gud	pero	lin
barenaked ladies	wayans	soledad	philippines
mst3k	dickey	espanol	noodle soup
truckers	isley	amor	malaysian
jethro tull	interracial	muy	for my next meal
canoe	nigeria	reggaeton	gangnam style

* The algorithm converted all words to lowercase and so I present them like that here.

Phish might've already given it away, but inside the white man rages a music festival for lumberjacks.

As for the other three lists, I had never heard of Zane or Anthony Hamilton or *The Coldest Winter Ever* or *Chungking Express* or Dash Berlin or a lot of the above before my scripts coughed them up, and I'm not going to pretend that a few minutes with Wikipedia can stand in for an understanding of a culture. These are users speaking in their own voice, and I'm going to let them do just that, but I will point out a few broad trends: white people differentiate themselves mostly by their hair and eyes, Asians by their country of origin, Latinos by their music. But because of the way the math is set up, the three non-white lists are evidence of cultures that I, as a white man, am not *supposed* to know. Of course, we're all familiar with Spike Lee and Beijing and Shanghai, but these lists give us the "insiders'" view of a culture. It's stuff an outsider can't get from autocomplete, or in any other top-down way, because you can't wonder at what you don't realize is out there. "Why do Asian people like *Norwegian Wood?*" isn't a stereotype because not enough non-Asians are familiar with the book (by Haruki Murakami) and movie. I thought it was just a Beatles song, and if before this chapter someone had asked me if I'd seen *Norwegian Wood,* I'd have said, "I don't think they made videos back then." The lists above are our shibboleths. As such, they are something no one could generate a priori, by typing things into Google Trends or by searching millions of hashtags. Sometimes, it takes a blind algorithm to really see the data.

Here are the lists for women. As you can see, they're very similar in spirit to the male. Maybe a few more ballads.

most typical words for . . .

white women	black women	Asian women	Latinas
my blue eyes	soca	taiwan	latina
red hair and	eric jerome dickey	tall for an asian	colombian
blonde hair and	haitian	philippines	una
love to be outside	imitation of life	taiwanese	cumbia

white women	black women	Asian women	Latinas
mudding	zane	beijing	banda
campfire	coldest winter ever	coz	tejano
four wheeling	nigerian	boba	merengue bachata
phish	interracial	filipina	gusta
hunting fishing	rb and gospel	cantonese	puertorican
campfires	five heartbeats	asians	colombia
green eyes and	anita baker	wong kar wai	mana
redneck	crooklyn	shanghai	vida
auburn	neosoul	seoul	bachata merengue
ride horses	octavia butler	macarons	amor
old crow medicine show	housewives of atlanta	viet	musica
grateful dead	luther vandross	kimchi	english and spanish
mountain goats	zora	for my next meal	espanol
love country music but	waiting to exhale	singapore	salsa merengue
gillian welch	anthony hamilton	malaysian	todo
country girl	chrisette	hk	por
christmas vacation	locs	malaysia	mariachi
bill bryson	outside my race	noodle soup	marc anthony
riding horses	kem	cambodian	espa ol
eric church	octavia	norwegian wood	novelas
barn	real housewives of atlanta	hong kong	como
allman	calypso	chungking express	pero
willie nelson	know why the caged	rachmaninoff	venezuela
harley	did i get married	southeast asia	soledad
brunette	spike lee	vienna	mas
flogging molly	braxton	mandarin	tacuba

I discovered in the course of working with it that the algorithm we used to make these lists is flexible. You can just as easily run the math in reverse. This gives you the *antitheses* of a group—the stuff they especially don't talk about—which can be as illuminating as what they especially do. On the following page are the lists for the men; they are printed on a darker background to visually emphasize that these lists are the opposite of the previous ones. They are the words least used by these groups yet most used by everyone else, the negative space in our verbal Rorschach. The lists are worth reading all the way through:

most antithetical words for . . .

white men	black men	Asian men	Latinos
slow jams	borges	sence	southern accent
trey songz	social distortion	layed	from the midwest
robin thicke	tallest man on earth	layed back	ann arbor
smh	gaslight anthem	sence of humor	midwestern
musiq	snorkeling	truck driver	gumbo
merengue	belle and sebastian	6'4	freakanomics
laker	xkcd	realy	equity
ig	diet coke	anything else you wanna	discworld
kevin hart	surfboard	like what u see	shanghai
raised in nyc	totoro	and my son	scallops
hip hop rap rb	magnetic fields	u like what u	slopes
kpop	gogol bordello	care of my kids	university of michigan
george lopez	dropkick murphys	makeing	assessment
neo soul	rebelution	welder	parentheses
rb and hip hop	peru	hunting fishing	snowboarder
neyo	horrible's sing along blog	care of my son	nyt
knw	wakeboarding	wanna know anything else	dominion
gud	herzog	else you wanna know	msu
follow me	my blue eyes	raising my son	ellipses
jordans	guitar and sing	ask and ill	maple
handball	dr horrible's sing along	comedys	nigerian
soulchild	coachella	dnt	kenya
ne yo	dr horrible's sing	woman who wants	john irving
bachata	yo la tengo	i'm a single father	over a decade
basketball	airborne toxic event	somthing	cheesesteaks
paid in full	yosemite	careing	wall street journal
mos def talib	feynman	writting	alternatively
mangas	coppola	and my daughter	mistborn
abt	wind up bird	haveing	weber
utada	kar	brown hair	gravitate toward

The opposite-of-Latino list I found most surprising. Hispanic and white identities are often conflated by demographers; for example, the US Census has struggled for years to separate one from the other. But they can only use checkboxes on paper. Latinos' "most typical" list above and their "opposite" one here define the extremes. That first gives you the furthest reaches of Latin culture (music and language) and this second gives the "corn-fed" Midwestern white stereotype, which is one of the few white subcultures with no Latin influence. Also, please notice that the "least Asian" things are all misspellings, working-class occupations, and other underachievements, like single fatherhood. And of course there's "6'4."

The women's lists are equally rich, and I again suggest you take in every word. There's the awesome *my name is Ashley* in the Asian antitheses. And I have to say, as a point of professional pride—when you ask an algorithm "What *aren't* black women talking about" and it tells you "tanning," you know you did something right.

most antithetical words for . . .

white women	black women	Asian women	Latinas
filipino	belle and sebastian	bbw	midwestern
neo soul	tanning	god my children	cincinnati
musiq	bruins	single mother of two	classically
slow jams	tahoe	grandson	kenya
rich dad poor dad	simon and garfunkel	god my daughter	neal
corinne bailey rae	magnetic fields	mother of three	shanghai
bailey rae	sf giants	human services	financial services
salsa bachata	flogging molly	degree in criminal justice	classically trained
aaliyah	head and the heart	single mom of two	southern belle
jpop	dodgers	notice my eyes and	cutting for stone
smh	wavy	wanna know just ask	in new england
salsa merengue	naked and famous	mexican and chinese	antarctica
nujabes	social distortion	they are my world	kavalier
48 laws of power	mountain biking	being the best mom	full disclosure
musiq soulchild	portugal. the man	raising my children	gravitate toward
neyo	camera obscura	a better life for	brussels
2ne1	rancid	associates degree in	toronto
esperanza	yo la tengo	curly hair and	march madness
mangas	paddle boarding	madea	cambridge
zane	armin	im a single mom	adventures of kavalier
n.e.r.d	santa cruz	mexican and italian food	creole
coldest winter ever	ecuador	i'm a country girl	meetup
mines	ccr	ellen hopkins	parentheses
ratchet	the dog park	people notice my eyes	arbor

white women	black women	Asian women	Latinas
aventura	bbqing	my name is ashley	curl up with a
malcolm x	origami	brittany	for my next meal
asians	handshake	at a daycare	singer songwriters
carne	gabriela	my family my cell	ann arbor
hw	line is it anyway	want a man that	raleigh
earphones	sunblock	me and my son	interpreter of maladies

I've talked about race a lot so far, and I've done so, as I've said, because it's something rarely addressed analytically. And the data I have is ideal for tackling taboos. But sex is the single-most important grouping that humanity has. It's existed forever, even stretching back to when we were just one people, and perhaps because of those deep-time roots, gender roles are more universal and more stubborn than any other. It's easy to forget, given how ineradicable the color line can seem, that ideas of race are a product of time and place. The Irish and eastern Europeans weren't considered "white" until the 1900s; in Mexico, the indigenous Mayans and the mestizos with Spanish blood have been distinct ethnic groups (and political opponents) for centuries. Yet to most people from the United States, they're both just "Hispanic." But sexual division is a given in human culture—every culture, every time.

Paradoxically, OkCupid isn't the best place to explore the differences between men and women, at least through the method we've developed here. Your sex is built into how you use a dating site, so, for example, the most salient thing you find about (straight) women from their profile text is that they're looking for men, and so on. Sex and profile text are inextricable, and analysis gets you little more than tautologies. The ideal source for analyzing gender difference is instead one where a user's gender is nominally irrelevant, where it doesn't matter if the person is a man or woman. I chose Twitter as that neutral ground. The lists below were made using the same math as the OkCupid lists above, but they use the text from users' tweets.

most typical words for . . .

men	women
good bro	my nails done
ps4	my sissy
james harden	mani pedi
mark sanchez	my makeup
my beard	my purse
cp3	girls night
in 2k	my hair for
bynum	prom dress
the squad	girls day
bro we	retail therapy
manziel	thanks girl
in nba	my future husband
year deal	to dye
iverson	dress shopping
yeah bro	too girl
kyrie	happy girl
hoopin	bobby pins
free agent	wanelo
tim duncan	my boyfriend and
scorer	my belly button
offseason	my roomie
hof	girlies
xbox one	dying my
david stern	cute texts
yds	girl crush
fantasy team	my boyfriends
gameplay	eyebrows done
gasol	curl my
lbj	my hubby
bro u	us girls

This gives you the distilled essences of men and women—read and grow stupider. Remember, before you get depressed, that the method is *designed* to find what's unique about each group, find the things they don't have in common and bring them to the fore. It's the mathematical version of the guy at the state fair: caricature by algorithm instead of airbrush.

These are the words at the extremes, but for men and women, as for the ethnic groups before, the essential vocabulary ("the," "pizza," and so on) is shared. In fact, there's a growing consensus among psychologists that men and women are fundamentally very similar, despite the popular cosmology that has them on different planets. Researchers at the University of Rochester recently pronounced "Men Are from ~~Mars~~ Earth, Women Are from ~~Venus~~ Earth," concluding:

> From empathy and sexuality to science inclination and extroversion, statistical analysis of 122 different characteristics involving 13,301 individuals shows that men and women, by and large, do not fall into different groups.

And yet, though my method is built to tease out differences, it's hard to imagine two more opposite sets of interests than the ones listed above. I can't tell which side to root for here—on the one hand, it's surely a worse world where women fixate on their appearance and men live the beef jerky lifestyle. On the other hand, if men and women were exactly alike, life wouldn't be much fun. Same goes for the by-race lists above. Cultural differences, even if they're occasionally laughable, make the world a richer place.

The Mars/Venus thing, metaphor though it is, reminds me that the heavens are an ancient reference point for science. Aristotle looked to the emptiness overhead to verify his *aether*. Newton confirmed his law of inverse squares through the motion of Mars. Even Einstein wasn't truly Einstein until the sun and moon said so, in a 1919 eclipse that confirmed the theory of General Relativity. Even though we're working on nothing so grand as all that here, I have to say I hope that paper's snarky strikeout typeface is premature, at least for the things we like and talk about and the ways we spend our time. Look at it this way: if there were no planet out there but Earth, it would be a very boring universe.

Ever Fallen in Love?

A few years ago a couple of MIT students, as a class project, used Facebook's data to create a working "gaydar." It was a simple piece of software that behaved a lot like any human trying to make an educated guess about somebody: it looked at who the person's friends were. The program quickly learned to recognize that a certain balance of gays and straights in a guy's social circle reliably indicated his sexuality; it didn't need to know anything directly about him at all. As the *Boston Globe* put it at the time, "People may be effectively 'outing' themselves just by the virtual company they keep." After the students had trained it on known profiles, the software was able to correctly predict if a man was gay 78 percent of the time, just from the nature of his social graph. That's a highly robust result when you consider that the expected success rate, if the program were just guessing blind, would've been only . . . uh, like . . . 10 percent? 2 percent? 8? $\pi/2$?

That's just the thing—part of the reason the kids made a program to guess in the first place—nobody really knows how many gay people there are. Past estimates vary wildly, as past estimates are wont to do.* *The Kinsey Report* in 1948 was one of the first scientific attempts to get a real number; it drew many brows together over horn-rimmed glasses by suggesting that 10 percent of men and 6 percent of women were gay. Later studies, many politically motivated and all using either survey data or contrived setups in laboratories, have put the number as low as 1 percent and as high as 15.† We are now able to get a better guess by a different route, and improving the accuracy here is important because, as one study blandly put it, "This work can usefully inform public policy." All but four presidential elections since 1952 would've flipped had 5 percent of the electorate changed their minds, so the question of whether a group makes up 1 percent or 5 percent or 10 percent of the country is of primal interest to the political calculus. Although the number of gay people carries no moral weight—even if there were just one in the whole United States, he or she would deserve the same

* Please see a map of the world circa 1491 for more information.

† Survey data is frequently polluted by outside factors, like how the researcher chooses to word the questions or chooses to weigh sexual experience against sexual identification.

rights as everyone else—it's a simple practical reality that policy decisions depend on the actual size of the population.

Also, for a group historically so stigmatized, a well-supported number speaks up where the individual cannot. It says: I am here. Gay people are a somewhat unusual minority, in that they can seem straight, at least superficially, if they decide they must. This surely involves a painful choice between self-preservation and self-expression that few other people ever have to weigh. But aside from the clear cost to the individual, "the closet" costs our society, too, as secrecy allows old attitudes to go unchallenged—and prejudice unchallenged is prejudice perpetuated. By forcing people to hide, intolerance creates its own cynical logic: when a large portion of a group goes unrecognized, it only makes marginalizing the whole easier. Visibility, on the other hand, creates acceptance. Even at lower estimates, homosexuality is no more unusual than naturally blond hair—which something like 2 percent of humanity is born with. In fact, being gay appears to be much more common than that. It's just less accepted and therefore much more often forced from view. Think about that the next time you pick up a celebrity magazine.

Turning to the data, Google Trends again shows its power to reveal what people feel they cannot say. According to Stephens-Davidowitz, the Google researcher, 5 percent of searches for porn in the United States are looking for what he calls "depictions of gay men"—that's a catchall that includes straightforward queries like "gay porn" and related searches like "rocket tube," a popular gay portal. What's more, that 1 in 20 ratio is consistent from state to state, meaning that same-sex desire is unaffected by a man's political and religious milieu. This evenness has a few powerful implications. First, it frustrates the argument that homosexuality is anything but genetic. If men from such different environments as Mississippi and Massachusetts are looking for gay porn at equal rates, that's strong evidence that supposed external forces have little effect on same-sex attraction.

The second implication of the state-by-state sameness in the data—that is, what it reveals not so much about gay people but about intolerance—needs a little time to unfurl. In early 2013, when he was still covering politics for the *Times,* Nate Silver applied his famous poll-modeling technique to same-sex-marriage ballot initiatives across the country. As he had done in the presidential elections, he aggregated data to get a snapshot of public opinion in each state, and then he performed some forward-looking analysis to guess how those attitudes might evolve. Silver estimated that gay marriage will be legal in forty-four states by 2020.

An interesting thing about Silver's work on the question, which was based on political polls, is how it relates to another data source: what people in each state told Gallup about their own sexuality. Here are those self-reported numbers graphed against Silver's most current projections for the acceptance of gay marriage, state-by-state. I've coded each state by its legal treatment of gay marriage and labeled a few of the outliers, as well.

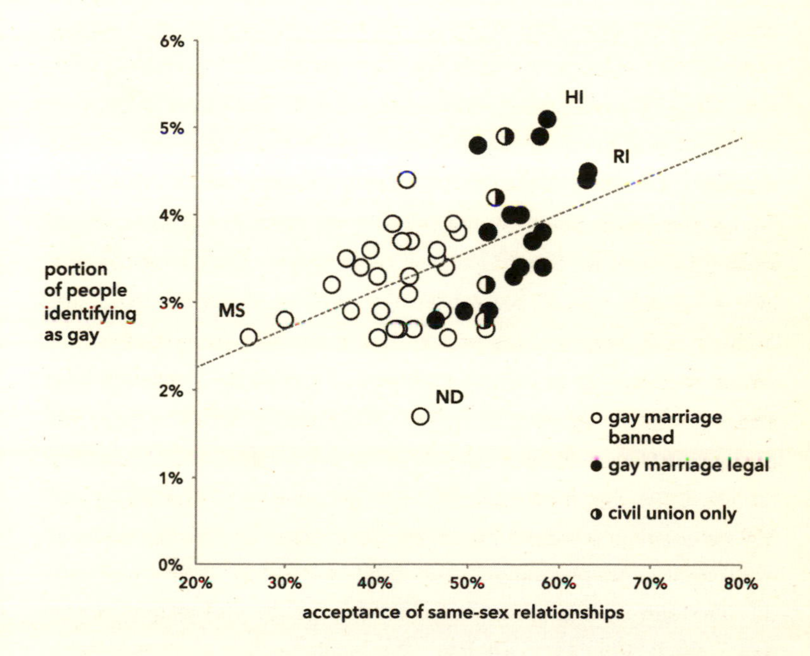

On the horizontal, you see that, per Silver, Mississippi is the least tolerant state and Rhode Island is the most. On the vertical axis, Gallup's numbers range from 1.7 percent in North Dakota, to 5.1 percent in Hawaii. And, as you see from the slant of the trend line, the more accepting a state is of homosexuality, the higher its self-reported gay population. Remarkably, if you walk that dotted line out to 100 percent support of gay marriage (statistically imagining a future world of perfect tolerance), you find it implies that roughly 5 percent of the population would say they are gay, absent social pressure not to be. That's the same number implied by Google Search, where the lack of social pressure isn't just theoretical.

Furthermore, that trend line isn't a function of folks simply living where they're more welcome. The state-to-state steadiness in searches for gay porn provides evidence of this and so does mobility data from Facebook. Comparing the hometowns of gay users to their current residences you find that relocation explains only a small fraction of the variance in Gallup's rates of homosexuality above. Gay people do not disproportionately move to more tolerant places. On the one hand, this is a testament to the strength of home ties, upbringing, and simple inertia. On the other, it means that for every person picking up and moving to a San Francisco or a New York City to live life fully, there are likely dozens still living in self-negation.

If you accept these two independent estimates of 5 percent, arrived at using three of the biggest forces in modern data—Nate Silver, Google, and Facebook, with an assist from that standby of old-school polling, Gallup—you begin to see those self-reported numbers in a different light. When Gallup tells us that, for example, 1.7 percent of North Dakotans are gay, then perhaps something like 3.3 percent of the state is gay and unwilling to acknowledge it. In New York, about 4 percent of the population is openly gay, leaving maybe 1 percent gay and silent. And likewise for every state. Against the steadiness of the data, the ups and downs in self-reported gay populations take on a new meaning: it shows a nation of Americans leading secret lives. This adds specific wisdom to the broad poetry often attributed to Thoreau: "most men lead lives of quiet desperation and go to the grave with the song still in them." These are refugees of the soul, and we see it in the data.

Data even gives us a picture of the collateral damage. Here's Stephens-Davidowitz again:

> In the United States, of all Google searches that begin "Is my husband . . . ," the most common word to follow is "gay." "Gay" is 10 percent more common in such searches than the second-place word, "cheating." It is 8 times more common than "an alcoholic" and 10 times more common than "depressed."

And those questioning searches are most common where repression is at its highest: South Carolina and Louisiana, for example, have the highest rates, and acceptance of gay marriage is below the national average in 21 of the 25 states where this search is most frequent. One wonders what the people so intent on driving homosexuality underground (or "curing" it) make of this data, and of the sexless marriages and children with unhappy parents their efforts so clearly create. Again, this isn't rhetoric—it's numbers. The old economic "misery index" is inflation + unemployment. I suggest the social version is the fraction of the population living in places where they can't be themselves. It's a situation that serves no end but suffering.*

Unfortunately, Google Search is ineffective for estimating the number of lesbians in the country. The many straight men looking for women-with-women porn garbles the data. However, we can see shadows of Silver's acceptance estimates in OkCupid's data, with some interesting twists. I estimate that more than a quarter of the country's dating gay population used OkCupid in 2013.† Gay online daters generally should

---

* And the political, religious, and entertainment careers of the people who perpetuate it.

† This is based on two assumptions: (1) that roughly 5 percent of the country is gay and (2) that, of the Census-reported 93 million singles in the United States, half are actually dating.

The government counts everyone who's not married as "single," which is obviously problematic in estimating the true single population, especially among gay people. In 2013, OkCupid recorded activity from 650,000 distinct gay profiles, which, by this arithmetic, is 26.8 percent of the actively dating American gay population. Some small fraction of the accounts are duplicates or "ghosts" (seldom used), but nonetheless the site's share of the country's gay dating market is substantial. In this note, as everywhere in this chapter, "gay" and "bisexual" users are counted separately, and this calculation does not include the latter.

be more open than average about their sexuality—after all, they're putting up profiles on a website. However, recognizing that many people would rather not broadcast their sexual identity Internet-wide, OkCupid gives its gay users the option to "hide" their profile from everyone except other gay users. Fifty-nine percent of gay men and 53 percent of gay women take advantage of the option. In this data too, the correlation between a state's tolerance and openness is visible, though more so for women, whom I've plotted below.

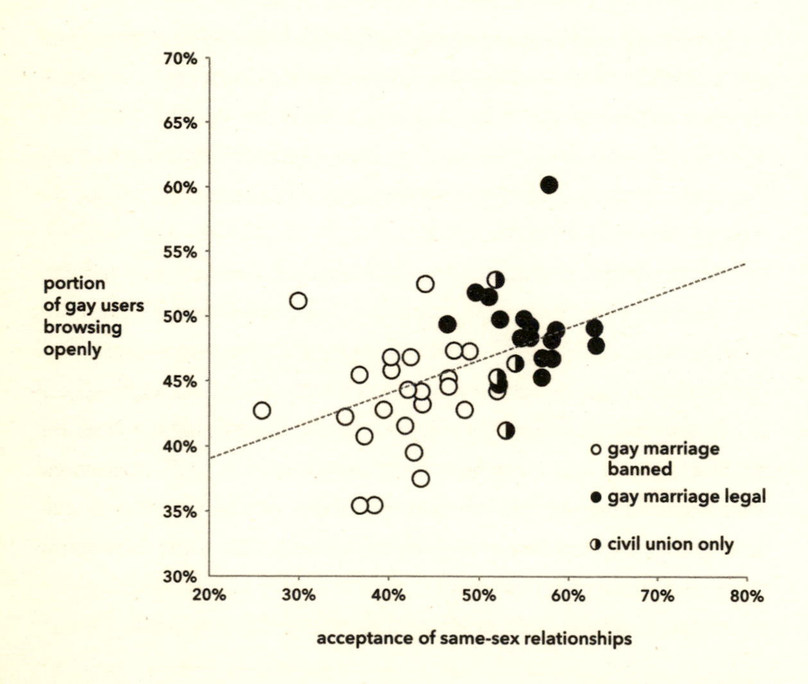

After you get past questions of "outness," gay users look a lot like everyone else on OkCupid. In the match questions, the site's gay users show the same rates of drug use, racial prejudice, and horniness as the straights, and gays want the same types of relationships. In fact, for sexual attitudes, if any group is an outlier, it's straight women. They're comparative prudes: 6.1 percent of straight men, 6.9 percent of gay men, and 7.0 percent of lesbians are on OkCupid explicitly looking for casual sex. Only 0.8 percent of straight women are, which probably says more about the taboo against sexual forwardness in (straight) females than anything else.*

The number of reported lifetime sex partners among all four groups is essentially the same. The median for gay men and straight women is four; for lesbians and straight men it's five, but just barely.† If there is a significant difference in sexual behavior, it's at the extreme end: there we find a stereotype partially fulfilled. Highly promiscuous gay men (the cohort reporting twenty-five or more partners) outnumber their straight male counterparts 2 to 1. Funnily enough, in sex, as in wealth and language, we have an inequality problem. According to this data, the top 2 percent of gay men are having about 28 percent of the total gay sex.

To see how identities are formed around the labels "gay" and "straight," we can apply the "word rank square" method from the last chapter to investigate personal self-descriptions. As before, profile essays give us a sense of what makes each group unique versus the others: what's special about lesbians, what makes gay men different from straight, and so on, and the method puts everything in the users' own words. The behavioral data above shows that *how* we love isn't all that different, but on the following page we see that *who* we love, of course, is. The math forces up the vocabulary most typical of each group:

---

* There are gay hookup apps specifically for casual sex: Grindr and Scruff are the best known services for men. The straight analogue for these apps is Tinder. It's proportionately as popular, perhaps more so. Therefore, I don't think selection bias (for long-term relationships) in OkCupid's gay population is any worse than in its straight population, though I do admit this is an impossible thing to know for sure.

† Forty-nine percent of straight men and gay women have reported four or fewer partners.

most typical words for . . .

gay men	gay women	straight men	straight women
first wives	i am gay	knows what she wants	honest man
velvet rage	old lesbian	i have no kids	man to share
tales of the city	i'm a lesbian	treat a woman	to meet a man
you're a nice guy	i am a lesbian	care of herself	a man who knows
anything on bravo	femme side	never been married	care of himself
music madonna	attracted to women who	daughter family	meet a man who
music britney	lesbian friends	for a good woman	find a man who
ltr oriented	are femme	treat a lady	who knows what he
romy and michelle's	butch femme	good women	meet a man
new guys	lesbian movies	my kids my family	man who knows how
barefoot contessa	single lesbian	hello ladies	a nice guy who
kathy griffin	u haul	type of girl	honest guy
single gay	butch but	woman that can	a man who has
the comeback	are feminine	real woman	are a nice guy
hiv positive	femme who	my son family	christian man
density of souls	elena undone	woman to share	like a man who
modern family glee	the butch	my daughter family	a guy who has
ab fab	not butch	intelligent woman	man that knows
most gay	movies imagine	god my kids	love jesus
muriel's	music brandi	girl that i can	a man who will
christopher rice	walls could	meet a woman who	man that has
muriel's wedding	lesbian romance	have no children	true gentleman
other gay	femme women	son family	you are a gentleman
flipping out	debs	with the right woman	guy to share
find mr	feminine women	treat her	nice guy who
guy to date	you're femme	right lady	like a guy who
sordid lives	soft butch	great woman	a guy that can
stereotypical gay	my future wife	a woman who can	christian woman
flight attendant	hunter valentine	nice woman	for a good guy
are you there vodka	lesbian looking	i like a woman	you're a gentleman

As before, I'll let you interpret the users' words in detail, and I'll just point out a few general trends. The two straight lists are all single-mindedly concerned with the person's (potential) partner. Every last entry for straight women is focused on the guy she's looking for (I'm counting Jesus here; he's single), and the men's only departure from talking about women is to note the presence or absence of children. These lists together read like "Me Tarzan, you Jane" in long form. Or maybe as adapted by Nicholas Sparks.

The lesbian list is more inward-looking, with more self-description, but it's still quite similar to the straight lists. Like straight women, lesbians are very much typified by the relationship they're looking for (*you're femme, my future wife*); they're just using different words.

The gay male list is very different from the other three. It's full of pop culture and has comparatively few references to the user's immediate person and family. *Anything on Bravo* has to be the most spot-on generalization of all time. That said, it's interesting that gay men are the least sex- and sexual identity–focused of all three groups. Or rather, they get their identity from something besides sex.

This method is, again, made to emphasize differences between the groups, but other data shows that the boundaries are porous. One of the most intriguing findings from OkCupid is the answer to this match question, asked only of the site's self-identified straight users.

Q: Have you ever had a sexual encounter with someone of the same sex?

	women		men	
Yes, and I enjoyed myself.	**22,308**	**26%**	**12,070**	**7%**
Yes, and I didn't enjoy myself.	**6,153**	**7%**	**10,100**	**6%**
No, but I would like to.	**14,896**	**17%**	**7,632**	**5%**
No, and I would never.	**42,286**	**49%**	**137,455**	**82%**
	85,643		**167,257**	

That is, 51 percent of women and 18 percent of men have had or would like to have a same-sex experience. Those numbers are far higher than any plausible estimate of the true gay population, so not only do we find that sexuality is more fluid than the categories a website can accommodate, we see that sex with someone of the same gender is relatively common, whether people consider it part of their identity or not.

The above data is from users who chose "straight" when signing up, but in that same pull-down menu OkCupid offers "bisexual" as an option. About 8 percent of women and 5 percent of men choose it. I have seen much frustration among bisexuals both on OkCupid and elsewhere with the idea that bisexuality is not a "real" orientation—that, for example, bisexual men are just gay men who haven't come to grips with it yet. Many gay people see bisexuality as a hedge. A recent study by the University of Pittsburgh Graduate School of Public Health puts it well, if a bit dryly: "Respondents who identified as gay or lesbian responded significantly less positively toward bisexuality... indicating that even within the sexual minority community, bisexuals face profound stigma."

Gerulf Rieger of the University of Essex, working with psychologists from Northwestern and Cornell, concluded in a 2005 paper that in terms of genital reaction to stimulus, almost all self-reported bisexual men were gay, some were straight, and very few were physically aroused by both sexes. He thus described male bisexuality as a "style" of interpreting arousal rather than arousal itself. Understandably, this infuriated the bisexual community; Rieger later revisited the topic to conclude that male bisexuality might be "a matter of curiosity"—that "interest in seeing others naked, observing someone else having sex, watching pornographic movies, or taking part in sex orgies" explained the apparent disconnect between bisexuals' self-reported attraction to both sexes and their observed physical attraction to only one. Their minds enjoyed all types of sex, but their bodies were more discriminating.

On OkCupid we find support for the spirit of Rieger's conclusions, if not his vague terminology. The vast majority of bisexual men and women seeks exclusively one sex or the other on the site. Below I've shown where the people who identify themselves as bisexual actually send their messages.

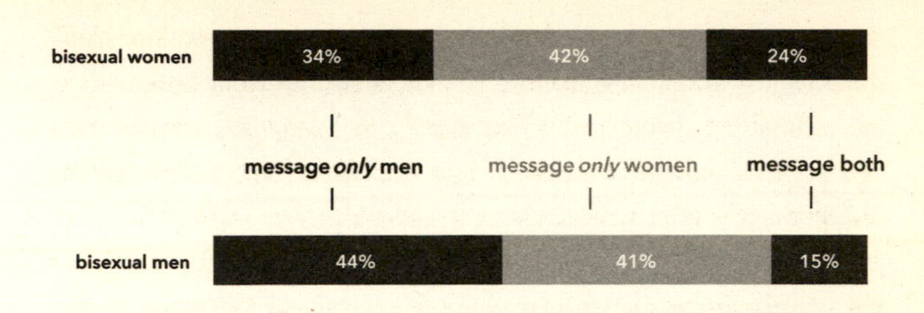

To land in either of the "message only" swaths, a user had to send 95 percent or more of his or her contacts to that sex, so the threshold there is quite high; this isn't an accounting trick. Only a fraction of the bisexual user base has any significant contact with both sexes. Whatever the mechanism, Rieger's claim that self-reported bisexuality doesn't reflect observed behavior appears correct in this case. Interestingly, for men, messaging changes over time. In that change we find plausible evidence for the hedge narrative: more than half of younger bisexual men message only other men, and that percentage drops steadily until the mid-thirties, at which point most of the male bisexual user base is messaging only women. This is what you would expect to see if men interested in men stop identifying themselves as bisexual as they get older and become more comfortable with being called "gay." But this question takes longitudinal data to fully answer, which we don't have yet.

That said, who we say we are and how we behave are two separate things, and the latter shouldn't automatically disqualify the former. People are ultimately free to describe themselves however they choose, and demanding that their labels fulfill a researcher's (or a website's) definition is pointless. Any discrepancy is ultimately the label's fault, anyhow—individuals love in whatever way feels right to them, and sometimes the words to describe it have to catch up. On Valentine's Day 2014, for example, Facebook launched more than fifty different gender options (allowing users to choose terms like *transgender* or *androgynous* instead of male or female). Ellyn Ruthstrom, president of the Bisexual Resource Center in Boston, was talking about orientation and Rieger's work, but could have been speaking to my data too, when she told the *Times*, "This unfortunately reduces sexuality and relationships to just sexual stimulation. Researchers want to fit bi attraction into a little box—you have to be exactly the same, attracted to men and women, and you're bisexual. That's nonsense. What I love is that people express their bisexuality in so many different ways."

We certainly find this varied expression when we look at the "typical" words in the profile text of bisexual men on OkCupid. In the top thirty are *bisexual, pansexual, cross-dressing,* and *heteroflexible*. In their antithetical list, you see *close with my family* and *really enjoy my job*—markers perhaps of the loneliness and disaffection that come from being an outsider, even among other outsiders.

Bisexuality for women is a bit different. It's more mainstream—or at least the version trafficked by the likes of Miley Cyrus is. Perhaps because marketers know that "sex sells" and that stars need to push boundaries, a kind of gay-for-pay lite is common in today's pop culture. In Miley's case—though of course I don't know for sure—it seems like a costume to sell records, no different from Gene Simmons's face paint. Similarly in costume, scammers targeting guys online will often select bisexual as the identity for their fake accounts. On Facebook, 58 percent of fake profiles are "female bisexuals" versus just 6 percent of non-fake. On OkCupid, the problem isn't quite that pronounced, but selecting bisexuality along with a few other key indicators guarantees you'll get special review from the site's admins.

But even on our legitimate profiles, which is almost all of them, female bisexuality and straight male fantasy are linked. You really pull this out of the data when you look at the profile text: it's mostly women inviting the world to threesomes with their boyfriends or husbands.

most typical words for . . .

bisexual women
bi female
bisexual female
me and my husband
me and my man
my boyfriend is

hubby and
we are a couple
i am bisexual and
me and my boyfriend
fun couple

---

couple we
married couple
we are not looking
fun with me and
do have a boyfriend
my bf and
female to join
girl to join
another couple
bi woman

---

my boyfriend my
i am bi sexual
my hubby and
join me and my
female for
my boyfriend and i
we are looking to
a triad
no single
send us

If I could put this to a beat and get Pitbull to do the middle eight, it would go straight to number one. That said, for all the crassness of sexual-identity-as-business-plan, it's a hopeful sign when a minority identity is something the mainstream thinks is worth co-opting instead of suppressing. Indeed, for sexuality, we see that things are changing, and quickly. Devising the projections we looked at above, Nate Silver clocked a marked change in American attitudes in the last decade. Acceptance of gay marriage accelerated markedly in 2004—and he determined, "One no longer needs to make optimistic assumptions to conclude that same-sex marriage supporters will probably soon constitute a national majority."

Thus, it all comes back to counting, and the fraction is going our way. Though people have been gay forever, in the late nineteenth century, people began to "self-disclose" their homosexuality as a political act. The phrase "coming out" was coined a few years later. Now, the goal of living and loving openly, which gay men and women have sought for so long, is near realized. The change is epitomized in the "out" celebrities, of course, but more so in the millions of other people whose names I'll never know but who have helped tick the metrics of acceptance ever so slightly upward. The day is coming when pollsters can put down their pens, scientists will turn their lenses another way, and enterprising students can use their algorithms to calculate other things. The day is coming when the world will be so open, no one will need to guess.

12. Know Your Place

When I was in junior high we had a long lunch period, and since everyone was too grown up at that age to really play or enjoy themselves, after the eating was over, we all just posted up outside the school and waited for the bell to ring us back to class. In the first few days of seventh grade, we sorted ourselves on the asphalt hardtop, and that arrangement, once set, hardly changed in three years. From nearest the cafeteria door to farthest, the order I remember is:

- ultra-coolest kids (mostly from the Heights, which was the wealthier part of town)
- the generically preppy kids
- the college radio REM/Cure people (this was pre-indie rock)
- the skaters
- the heshers (what we called the metalhead stoner types, and anyone else for whom glue was more than just an adhesive)
- me and my friends
- A BIG BROWN DUMPSTER
- exchange students and kids with learning disabilities

Obviously, this alignment was more than just random. The dumpster, god bless it, created a natural gathering point for the untouchables, and from there the +/– polarity of the student molecule took over. Given that at one end of the line my people were playing pencil-pop and debating the merits of Teenage Mutant Ninja Turtles, The Role Playing Game *Not* The TV Show Because The TV Show Is For Kids, everyone else fell into place by fundamental force.

One of the beautiful things about digital data, besides its sheer volume, is that, like the back lot at Pulaski Heights Junior High, it has both physical and social dimensions. A piece of paper has two axes, space-time four. String theory predicts that our physical existence requires somewhere between ten and twenty-six dimensions. Our emotional universe surely has that many and more. And in combining these spaces—our interior landscape with our external world—we can portray existence with a new depth.

The way we've looked at people and interaction so far—connections, profile text, ratings, and so on—has mostly ignored physical place, but websites and smartphones are of course gathering ample location data. Tweets are geotagged with latitude and longitude; Facebook asks for your hometown, your college town, your current home; many apps know the very building you're standing in. Here we're going to layer identity, emotion, behavior, and belief over our physical spaces and see what new understandings emerge. We'll look at how location shapes a person, and how people have laid new borders over our old earth.

The boundaries of many communities were created by fiat or accident—or both. The United States and the USSR split Korea on the 38th parallel because that line stood out on a map in an officer's *National Geographic*. Earlier that same month, Germany was divided into zones of occupation that reflected, more than anything else, whose troops were standing where at the time. Many of our own American states were created by royal charter or act of Congress, their borders drawn by people who would never see the land in person. Absentee mapmaking was and still is a much more pernicious problem in Africa, the Indian subcontinent, the Middle East—and everywhere else the tread of Empire has stamped the soil. Only very occasionally have maps been drawn to reflect "the will of the people," and even in those cases, as we've seen in Israel, which began its modern history as, officially, the British Mandate for Palestine, the question naturally becomes: which people, whose will.

For websites, political and natural borders are just another set of data points to consider. When information—fluid, unbounded, abstract—is your currency, the physical world with its many arbitrary limits is most often a nuisance. At OkCupid, rivers are an endless irritant to the distance-matching algorithms. Queens is both a half mile and a world away from Manhattan. Try explaining that to a computer. The problem is that when a person is online, he or she is both of the world and removed from it. But that duality also means we can remix our physical spaces along new lines, ones perhaps more meaningful than those drawn by plate tectonics or the dictates of some piece of parchment.

Here you see a plot of how Craigslist carves up the country—each region in the map is the territory served by a separate classified list. One mapmaker called it the "United States of Craigslist" but "united" feels to me like the wrong word—this is a partition, and, within the whole, each little zone is its own petty kingdom. It's a Holy Roman Empire of old furniture.

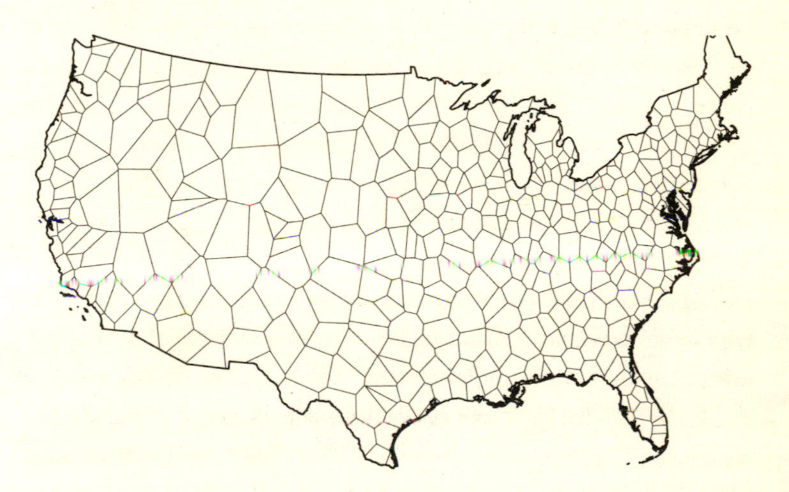

Once we begin to graft content to the spaces, the map becomes more interesting. On the following page is Craigslist's empire again, but overlaid with the most popular locations listed on the site's many "Missed Connections" board, where a lonelyheart might post something like:

> Both of us boarded the uptown Q at 34th. You were wearing a peacoat and your eyes had that Audrey Hepburn twinkle. We locked stares a few times; if you read this email me.

That's the Manhattanite's version, at least. Portlandia most often makes eyes on the bus. California flirts by the elliptical machines. But for much of the rest of the country, the venue of longing is Walmart.

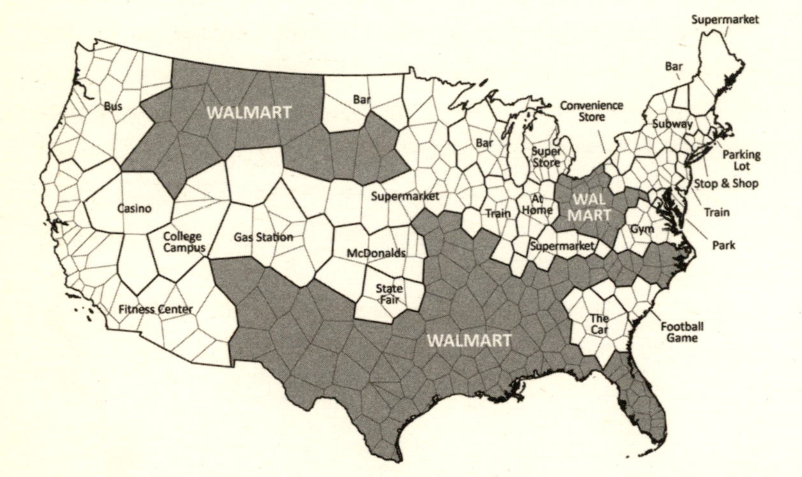

Now we're getting to a place that a traditional cartographer can't take us, that no satellite can pick up. The above is a simple and goofy page from a new kind of atlas: behavioral and physical terrain as one.

In the above examples, Craigslist defined its borders a priori, by picking the markets they wanted to serve. Most websites *collect* location data rather than project it, and from these we can create a truly alternate map of the world, actually move the borders and contours to fit the human landscape. Years ago, an enterprising hacker scraped data from Facebook and plotted the shared connections of the 210 million profiles he'd gathered. From the data he saw, he divided America into whimsical states defined by friendship rather than politics. There were seven of them—Pacifica (the Pacific Northwest), Socalistan (California), Mormonia, the Nomadic West, Greater Texas (which included Arkansas, Oklahoma, and Louisiana), Dixie in the Southeast, and then, in a bright green swath stretching from Minnesota down through Ohio and over to the Atlantic covering all of New England, Stayathomia. My kind of country.

Since then, smartphones, each one with a tiny GPS pinging, have revolutionized cartography. Matthew Zook, a geographer at the University of Kentucky, has partnered with data scientists there to create what they call the DOLLY Project (Digital OnLine Life and You)—it's a searchable repository of *every* geotagged tweet since December 2011, meaning Zook and his team have compiled billions of interrelated sentiments, each with a latitude and longitude attached. DOLLY is an incredibly versatile resource, the output of which is only now being explored. For Zook, it's already had a few highly personal applications. In February 2012, his office in Lexington was shaken by an earthquake, and he turned to the database to see the psychological aftershocks. The map below shows the density of reaction on Twitter, plotted over the physical epicenter of the fault. Here we see contours of surprise laid over the shifting earth:

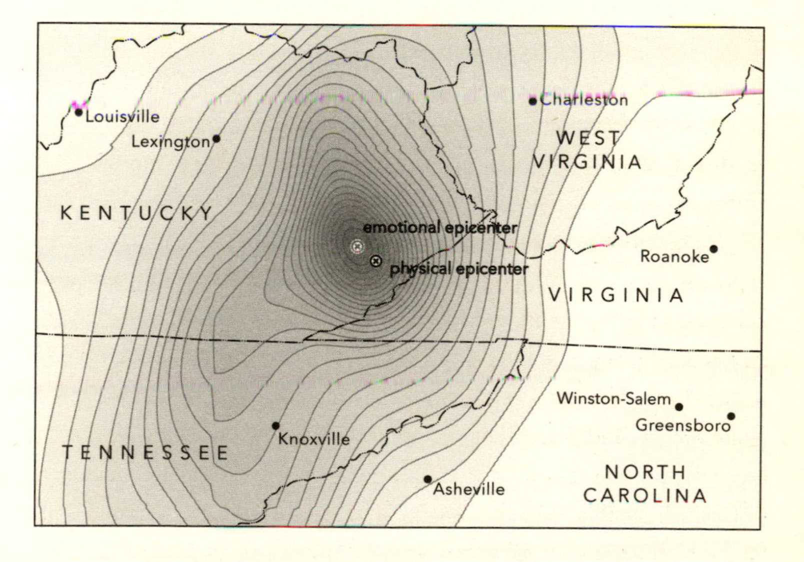

Zook discovered that the quake's emotional epicenter was just northwest of the seismic one, in Hazard, Kentucky, and as simple as it sounds, this kind of finding is truly new. The Craigslist maps, for example, could've been made in the 1970s—after all, the idea for the website's "Missed Connections" section was lifted from newspapers. So before the Internet, if you'd really wanted to, you could've clipped a month's worth of listings from the main daily in, say, each of the country's top 100 cities, logged the data, and gotten very close to what we saw a few pages ago. Even the Facebook/Stayathomia redefinition was theoretically possible decades ago, provided a research team had the resources to interview millions of people in their homes and track down their stated connections.

But Zook's map shows people's instantaneous reaction to an event that lasted a split second. Surveying Kentuckians later, even with infinite effort, he couldn't have generated a true report—not only do emotions change in the remembering, but media coverage and talk about the quake would've hopelessly polluted the data. People with smartphones don't make seismographs obsolete but Zook's plot reflects the "impact" of the earthquake in a much more direct way than the old Richter scale. Knowing nothing else about a quake, if it were your job to distribute aid to victims, the contours of the Twitter reaction would be a far better guide than the traditional shockwaves around an epicenter model.*

Even though each one is transitory, tweets collected together can capture more than ephemera. A demonstration of DOLLY's power on YouTube shows it tracking the Dutch holiday of Sint Maarten, a sort of Germanic Halloween where children go door to door singing for candy. In the data, you see people celebrating not only in the major population centers of the northern Netherlands, as you'd expect, but also in Western Belgium—the tweets reconnect old Holland to Flanders, its cultural

* Two months later Zook measured a convulsion of another kind: the Kentucky Wildcats won the NCAA championship and the students got wasted and burned shit like the future leaders they no doubt are. #LexingtonPoliceScanner began trending as a hashtag, based mostly on this tweet from @TKoppe22: "Uh We have a partially nude male with a propane tank #LexingtonPoliceScanner." Zook tracked that tag to show how formerly local nonsense can now reverberate worldwide. The highbrow/lowbrow schizophrenia of Twitter never stops amazing me. It's the Chris Farley of technologies.

cousin. Thus we watch an animated visualization of GPS-enabled data points, and see shadows of the Hapsburgs.

Given the power of what we can already see through software like DOLLY, the lack of longitudinal data is especially painful. On today's research corpus, time often feels like a phantom limb. Twitter currently gives us so much of that multidimensional promise: we have every emotion, we have every spot on the globe, but we still have only a few years to work with. In Europe, where the combination of geography, culture, and language has been so volatile over the centuries, imagine being able to track the Alsace-Lorraine as it changed hands—German, French, German, French—each government imposing its culture on the people, as if the region were a house taking on coats of paint. Or imagine the Caribbean basin in the late fifteenth century and being able to watch first the soldiers, then their religion, then their language overwhelm the land, Arawak to Aztec. To see the ebb and fracture of a culture over decades is what DOLLY was built for. All it needs now is the decades themselves.*

Geocultural insights can be found in other sources, too, and though in most of them you lose the immediacy of Twitter, you get a different kind of depth in its place. When websites pose questions directly to their users, we have a chance not only to refine borders but to show they don't really exist as normally conceived.

On the following page are one million answers to "Should burning the flag be illegal?" collected by OkCupid. Here my mapping software drew no political or natural boundaries, it just organized belief according to latitude and longitude. This is truly a nation defined by its principles, or, as you can see, *two* nations: Urban and Rural. You can even see where one encroaches on the other: the rural communities up the Hudson River and in Northern California's wine country, built up with Big City money, have Big City opinions as well.

---

* I realize an added condition is that the affected people use Twitter, and that in the context of pre-Columbian Mesoamerica that's an absurd expectation. However, as I've said before, the service is much more pervasive and more democratic than most people think, and if anything similar to the Spanish Conquest were to happen today, you most certainly would see the reverberations on Twitter.

map of answers to "Should burning the flag be illegal?"

Similarly, and in support of the earlier Google Trends finding that homosexuality is universal, we see that same-sex searches have no borders, no state, no country. Below is a plot of gay porn downloads, by IP address, taken from the largest torrent network, Pirate Bay. This map, too, is without any pre-drawn guides, and as opposed to the OkCupid plot above, its theme is solidarity: from Edmonton and Calgary down to Monterrey and Chihuahua, this is just where people live.

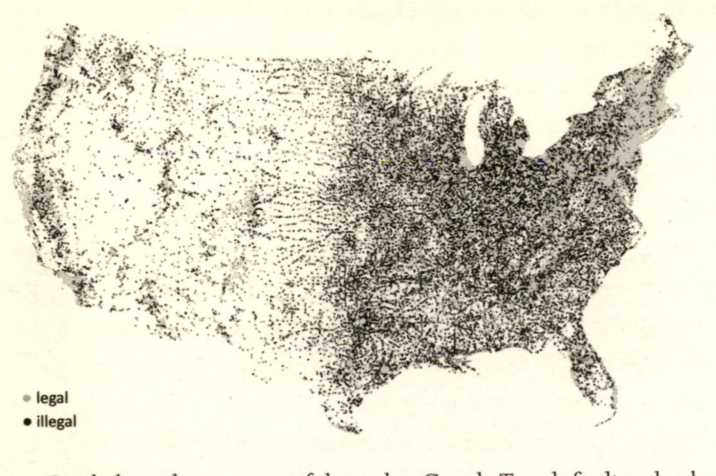

There are as many ways to draw maps as there are sources of data. We've been slowly working our way up off the page, building a psychological dimension—how we feel about the flag, porn—on top of our maps. But it's possible to go the other way: data can tie abstractions back down to earth. Take cleanliness, again via OkCupid. This is how often people say they shower:

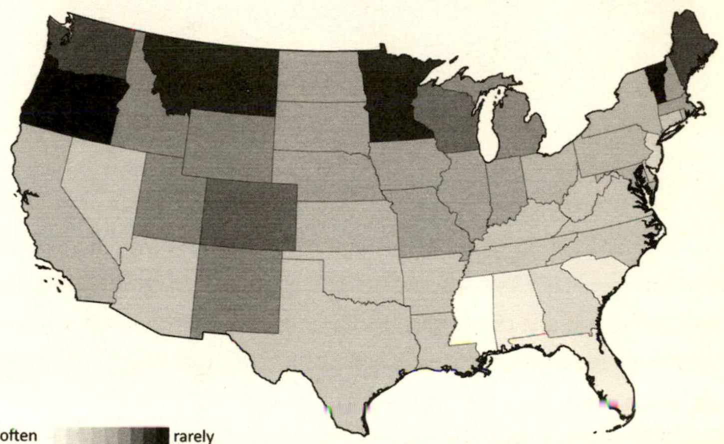

map of answers to "How often do you shower?"

On the one hand, the broad trend merely reflects the weather: where it's hot, people shower more. But down in the details there are a pair of good stories. In Jersey's lightness, you can read the gym/tan/laundry grooming obsession of Pauly D and the Situation—Jersey is much more fastidious than the surrounding states. And in Vermont you find the opposite philosophy: the crunchiness is more than just a stereotype. Vermont's the most unwashed state overall, and truly an outlier compared to its immediate neighbors. According to Google the state animal is the Morgan Horse. It should be a white guy with dreads.

Politics, weather, Walmart, and certainly earthquakes all have a strong connection to the physical world, but in some of our data we can begin to see an exclusively inner geography. Take lust, which in theory, should have no state. But here we see it does, and a surprising one:

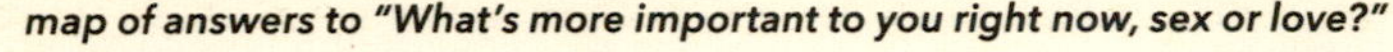
map of answers to "What's more important to you right now, sex or love?"

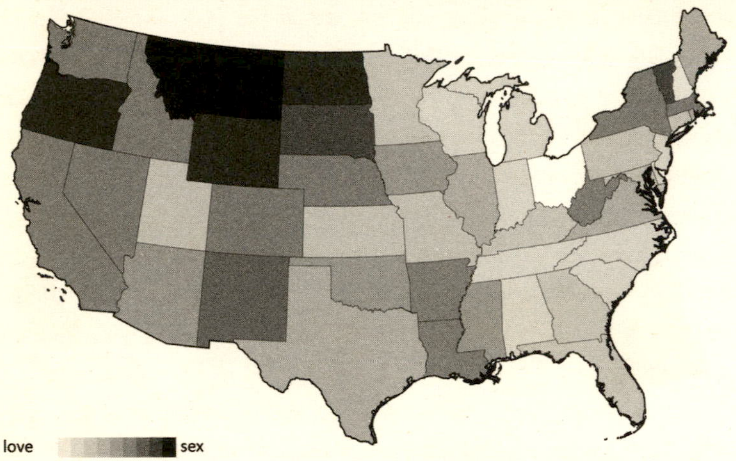

This pattern comes up again and again on OkCupid—the north central and west of the country is more sexually open, more sexually adventurous, and more sexually aggressive. Up the Pacific Coast you'd perhaps expect such unconventional attitudes, but for many of these red-meat states, it goes against type. Politically, OkCupid's users in, say, the Dakotas are as conservative as their reputation. Their profile text isn't much different from anyone else's. For all other indicators, the states should not be dark, but in the data we see a mysterious sexual intensification. This unexpected pattern reveals a further power in Internet data; we can now discover communities that *transcend* geography, rather than reflect it.

This data above does *not* prove that the Mountain Time Zone is one big high-plains makeout party. In fact, the explanation is rather banal: if you are looking for people to have sex with in a place like Pierre, South Dakota, your local options are limited. So you try a dating site to find what you want. It's simple selection bias in our data, but there's meaning there: where people can't find satisfaction in person, they create alternative digital communities. On a dating site, that means communities with similar sexual interests. On other sites with more diverse aims, where the users aren't just there to flirt in groups of two (and occasionally three), you get something richer.

Reddit is the fulfillment of that earliest ambition of the Internet—to bring far-flung people together to talk, debate, share, spread news, and laugh. To collapse space and create personal closeness. It's one of the most popular sites on the web,* and it rightly calls itself "The Front Page of the Internet"—a lot of the ridiculous viral stuff you see on the big aggregator sites originates there. There's a video trending on the *Huffington Post* as I write this—no joke—with the headline: "This Deer Thought No One Was Watching It Fart, Now the Whole World Knows." I promise you, Reddit was watching it fart *first*.

The odd thing is, for all its influence, Reddit doesn't really *do* anything; there are no apps, no games, no profiles to speak of. Their New York office is in a co-working space and smaller than my bedroom. The site itself is just a raw list of links submitted by the users, who vote, and comment, and comment on the comments, and modify, and repost all day long, in what feels like the world's biggest group of friends sitting on the world's longest couch. Few Redditors know each other's names, let alone ever meet in person, yet their bond is no less close for being anonymous: a forty-year-old woman in the Bay Area was alone the day before Thanksgiving 2011 and posted as much. Her thread received over 500 comments in just a few hours (including, of course, many invitations to the next day's dinner) and the post quickly broadened, completely ad hoc, to connect Redditors in many other cities.

The site is self-organized into thousands of themed subreddits. Each of those is user-created and -moderated, and each has its own devoted set of posters and commenters. These are places where people have created true virtual communities from nothing but wide open space. There's *gaming, technology, music, nfl,* alongside a lot of home-grown topics that you'll only find on Reddit:

* In December 2013 it had 101 million unique visitors and served 5 billion pages.

explainlikeimfive—an example post: "In Hinduism and Buddhism where the dead get reincarnated, how do they account for population growth?"

iama—"IamA reporter covering NJ Gov. Chris Christie. AMA! [ask me anything]"

todayilearned—"TIL that the town of Boring, Oregon has 'paired up' with the town of Dull, Scotland to promote tourism in both places."

askreddit—"Ex-smokers of Reddit, what ACTUALLY WORKED to get you to successfully stop smoking?"

whowouldwin—"Superman Prime vs Superman w/infinity gauntlet"

On the next page I've plotted the two hundred most popular topics, and this is something you could properly call "the United States of Reddit." It's a geography like the Craigslist division we saw before—made, in fact, by a similar algorithm—but instead of physical geography, it plots a geography of interests, of the collective Reddit psyche. And it shows distinct yet connected communities. The size of each state corresponds to the popularity of the topic, and the software put "like with like," according to cross-commenting between subreddits.

As we did before when we encountered an unfamiliar way to present verbal data, you should search out a few known terms to get a feel for how everything fits together. For me, this was easy. My favorite game, Magic: The Gathering (*magicTCG*), is correctly surrounded by its unfortunate natural friends *MensRights*, *whowouldwin*, and *mylittlepony*. Similarly, many sports (*nfl*, *nba*, *formula1*, and so on) are grouped at the bottom. Everything *pokemon* is clustered over to the left. *Britishproblems*, along the right edge, is next to *australia* and *soccer*. It also makes sense that the most popular subreddits are in the center—that is, not too far from anything. The gray tint corresponds to how tight-knit each subreddit is. It shows the degree to which the people posting post only there. The darker the gray, the more isolated the thread. This whole thing is an abstraction, but it shows how people can locate themselves by what they find interesting or funny or important rather than where they happen to sleep at night. It's a map of one particular collective consciousness.

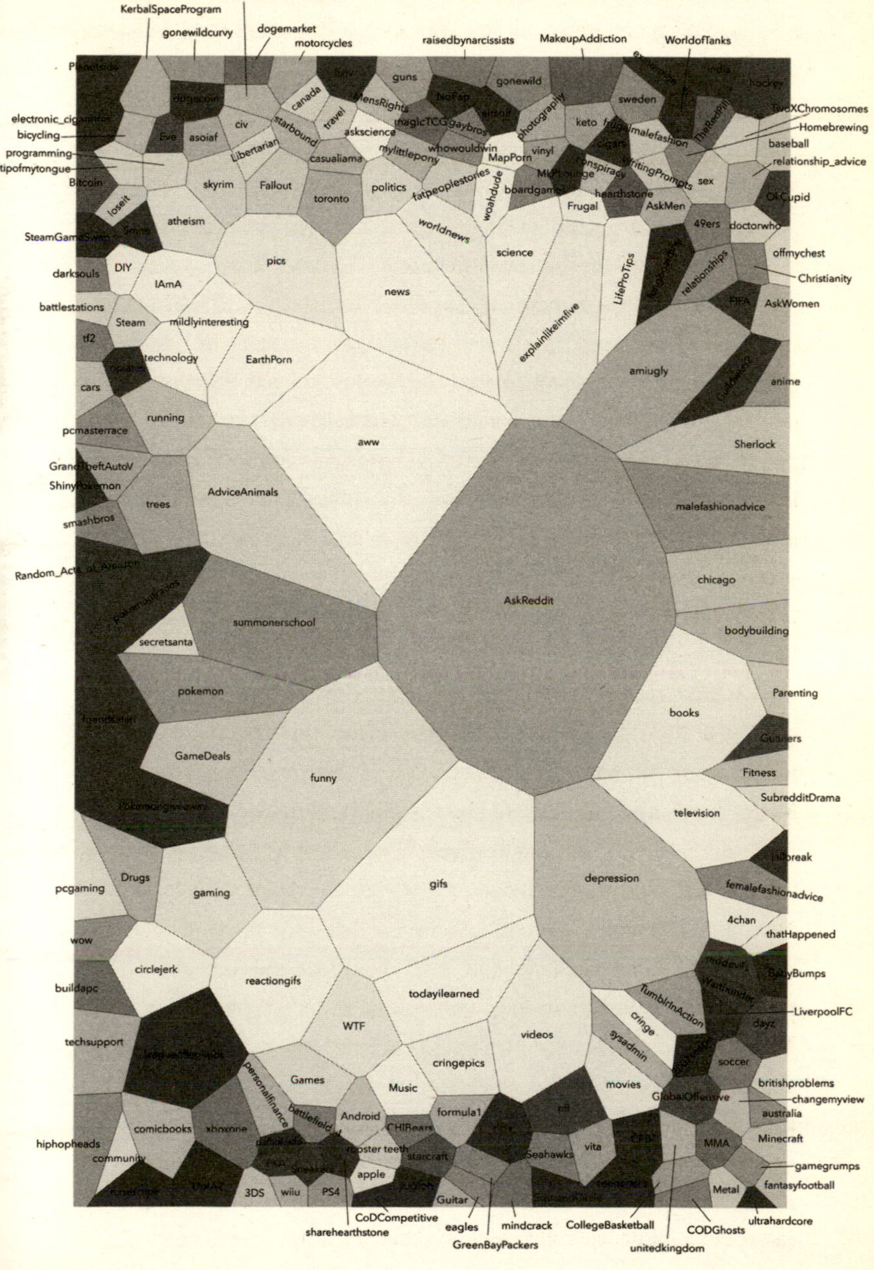

RandomActsOfGaming
KerbalSpaceProgram
gonewildcurvy
dogemarket
motorcycles
raisedbynarcissists
MakeupAddiction
WorldofTanks
electronic_cigarette
bicycling
programming
tipofmytongue
Bitcoin
SteamGameSwap
darksouls
battlestations
tf2
cars
pcmasterrace
GrandTheftAutoV
ShinyPokemon
smashbros
Random_Acts_of_Amazon
TwoXChromosomes
Homebrewing
baseball
relationship_advice
OkCupid
offmychest
Christianity
AskWomen
anime
Sherlock
malefashionadvice
chicago
bodybuilding
Parenting
Fitness
SubredditDrama
femalefashionadvice
thatHappened
BabyBumps
LiverpoolFC
britishproblems
changemyview
australia
Minecraft
gamegrumps
fantasyfootball
ultrahardcore
CODGhosts
unitedkingdom
CollegeBasketball
mindcrack
GreenBayPackers
eagles
CoDCompetitive
sharehearthstone
hiphopheads
pcgaming
wow
buildapc
techsupport
canada
guns
gonewild
sweden
hockey
MensRights
NoFap
travel
starbound
civ
Eve
asoiaf
Libertarian
askscience
magicTCG
gaybros
photography
keto
frugalmalefashion
TheRedPill
cigars
whowouldwin
mylittlepony
casualiama
vinyl
MapPorn
conspiracy
WritingPrompts
sex
skyrim
Fallout
toronto
politics
fatpeoplestories
woahdude
boardgames
hearthstone
AskMen
Frugal
49ers
doctorwho
loseit
atheism
pics
worldnews
science
news
LifeProTips
relationships
FIFA
DIY
IAmA
Steam
mildlyinteresting
technology
EarthPorn
explainlikeimfive
amiugly
running
aww
AdviceAnimals
trees
summonerschool
secretsanta
AskReddit
pokemon
GameDeals
funny
books
television
Drugs
gaming
gifs
depression
4chan
circlejerk
reactiongifs
todayilearned
WTF
videos
TumblrInAction
cringe
sysadmin
dayz
soccer
movies
GlobalOffensive
cringepics
Games
Music
personalfinance
battlefield
Android
formula1
CHIBears
nfl
vita
Seahawks
MMA
comicbooks
xboxone
rooster teeth
starcraft
community
apple
Metal
3DS
wiiu
PS4
Guitar

Benedict Anderson is a professor at Cornell University, and he wrote a book that sat unopened on my bookshelf a long time. I was supposed to read it for a college class and didn't, but through all my moves over the years I've carried it with me; it's been a stowaway in every U-Haul. The book's called *Imagined Communities*, and I opened it recently because the title finally seemed applicable. Anderson's main topics are nationalism and nation-building and he suggests that a nation "is imagined because the members of even the smallest nation will never know most of their fellow-members, meet them, or even hear of them, yet in the minds of each lives the image of their communion." He was writing in 1981, but he could have been talking about the Internet. I don't know if Reddit is a nation, but it's got plenty of communion. And it's interesting to see another purely digital community define its burgeoning identity. Earlier we saw the ancient rush to communal violence, as directed at Safiyyah, Natasha, and Justine on Twitter. Here, on Reddit, we see a few of nationhood's better angels: belonging, sympathy, sharing.

I've lived now in Brooklyn for twelve years—*Imagined Communities* had collected quite a bit of that New York City schmutz by the time I pulled it down to read—but the first place that book ever went with me was Texas. Right after school, I had been living with a few other guys, and one of them, Andrew Bujalski, who's now a director, decided to move to Austin because he loved *Dazed and Confused* and *Slackers*. He was making a pilgrimage to find Richard Linklater. The rest of us had no plan, so we just attached ourselves to his.

Of course picking up and moving like that is the privilege of twenty-two-year-olds with nothing better to do but chase someone else's dream. We'd heard Austin was cool, so we went there. It's a lightweight example, but group movements like this, based on little more than word of mouth and hope for something better, created the world as we know it. The Great Migration—millions of African Americans leaving the Jim Crow South for cities like Detroit, Chicago, and New York in the early 1900s—was a transformative cultural shift for the country and was made of thousands of small-scale pick-up-and-move decisions. Same with the gold rush that settled California. Same with much of the European settlement that brought the Old World to this continent in the first place. Same with, I imagine, the bands of Clovis people who crossed the ice bridge 13,000 years ago to become the very first nation on this soil. Communities move to find an environment that will sustain them and where they are safe, but also to find a physical place that reflects what they feel within.

Recently, Facebook's Data Science team took a worldwide look at modern large-scale movements—coordinated migrations, where a significant proportion of the population of one place has moved, *as a group*, somewhere else. People don't move en masse like this in the United States much anymore, but in many places, they're just beginning to. The researchers plotted coordinated movements around the globe. Here I've excerpted a small section of their map of Southeast Asia: the lines show small towns and villages relocating wholesale to urban centers. It's a static picture of a rapidly changing region. For what it's worth, this could've been England circa 1850, or the United States fifty years later.

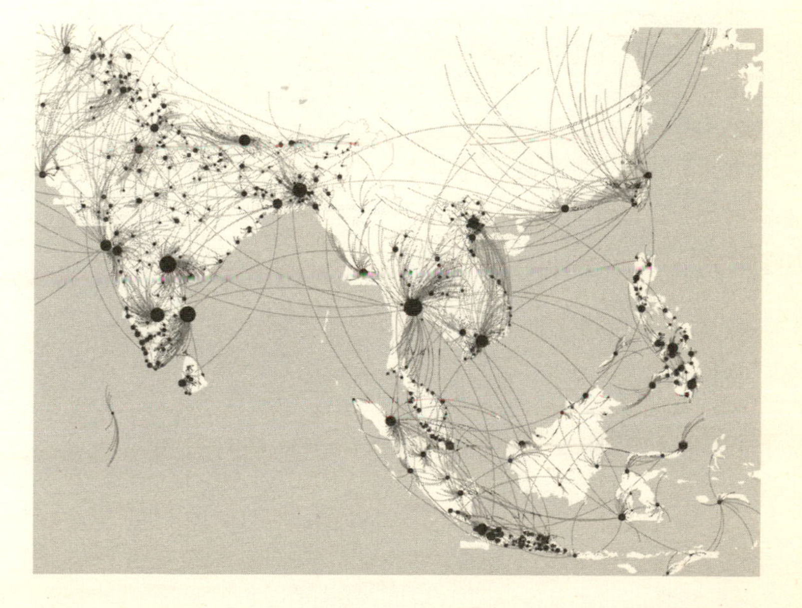

In the broadest sense, these moves are most likely driven by economics—cities like Chicago or Bangkok promise jobs. But though the lines and dots on this map are aggregates, the migrations they reflect are all small, personal, and, no doubt, unique to the people making them. Was it a parent who made the decision to pack up and go? Did a friend lead the way? Who did these people join in their new city? Who did they leave behind in the old? Did they bring everything? Leave everything? And I can't help but wonder, too, does everyone have a book that follows them until they read it? And, if so, what is theirs?

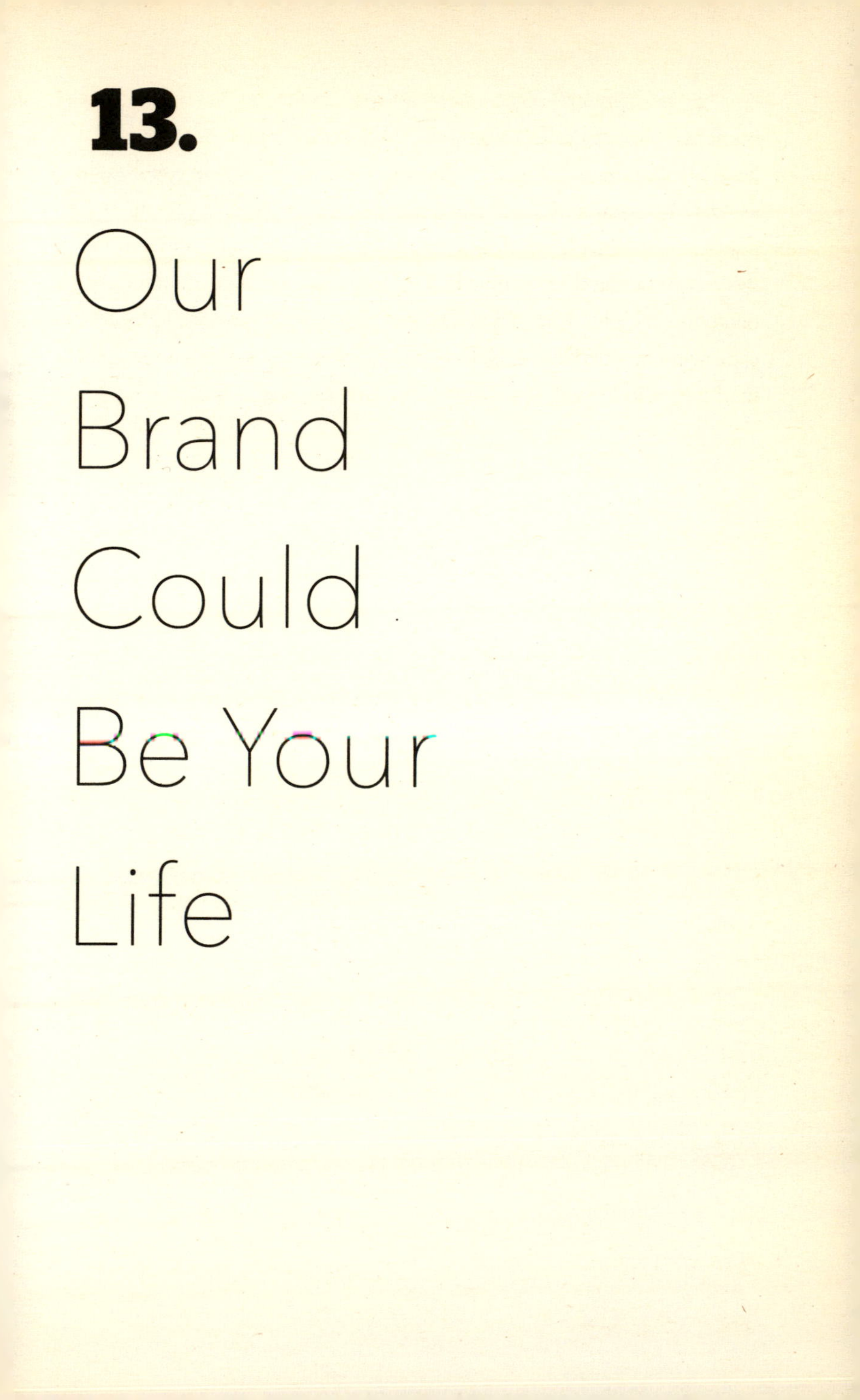

13.

Our Brand Could Be Your Life

Bass Ale's triangle logo was the first registered trademark in the English-speaking world, and today that sturdy oldness is a big part of the brand's appeal. They lay it down right there on the label—"England's first registered trademark." But what they don't tell you is that Bass was only first because a brewery employee happened to be first in the queue at the registrar's office the morning that Britain's Trademark Registration Act took effect. They've parlayed an accident of bureaucracy into a reputation that, at least judging by what's in those brown bottles today, far outstrips the actual quality of the product. Bass is a brand built on nothing more than the act of branding itself.

There were many brands and marks before Bass—enough for the UK to begin to regulate them, after all, and labels and image-making pre-date even the Industrial Revolution. I mean, brands were originally burned into flesh. It's hard to get more primitive than that. Archaeologists have unearthed branded oils and wine in desert tombs sealed five thousand years ago. One label found in Egypt reads "finest oil of Tjehenu" beneath the royal emblem and a pictograph of a golden oil press. Compare that to the "choicest hops, rice and best barley" beneath the "King of Beers" on a can of Budweiser—as far as branding has come, in many ways it will probably always be a Bronze Age science, because the emotions it plays to are eternal.

But while aspiration and the prestige of association may be timeless concepts, truly new territory has recently opened to the brand: people. In 1997, Tom Peters, a motivational speaker and management consultant, published an article called "The Brand Called You" in *Fast Company* magazine, and the era of personal branding was born.

His article, really more of a sales pitch, asks readers to first determine their "feature-benefit model" and then to relentlessly market it to employers, coworkers, and the larger world . . . or else! Those are literally the last two words, and they punctuate all the typical hokum ("Sit down and ask yourself . . . what do I want to be famous for? That's right—famous for!" and "You are a leader. You're leading You!") that the worst business writing has to offer. Reading it, you imagine Mr. Peters miked up and pacing the rostrum like a lion caged—caged by that darn paradigm that he's about to explode before your very eyes, with truth bombs, know-how, and exclamation points. He shows the kind of belief that a different type of person channels to rip phone books in half for his tight bro J.C. The byline at the bottom of the piece reads, "Tom Peters is the world's leading brand when it comes to writing, speaking, or thinking about the new economy." He was also, at that point, not just the leading, but the *only* person calling himself a brand. Hence a mouthpiece for the "new economy" takes a page from Bass's Victorian playbook. And why not? Fake it till you make it. The article kicked off the idea of self-branding as a direct path to success and is still read in marketing classes today.

A few years later, a man named Peter Montoya expanded upon Peters's idea in a second influential manifesto called *The Brand Called You.* Yes, it had the same title as the original manifesto, and no, he and Mr. Peters did not work together; in fact, if anything, the two men are rivals in the branding-guru business. Melding the cold steel of cluelessness to brass balls is the well-paid talent of pitchmen everywhere, and Mr. Montoya just might be the master wizard. *The Brand Called You* (his version) is essentially one long outline, and this is the very first bullet point, which appears on page 2:

> 1. You Are Different. Differentiation—the ability to be seen as new and original—is the most important aspect of Personal Branding.

Naturally, *The Brand Called You,* the remake, was a bestseller, and Montoya, like Peters, has a thriving speaking career to this day. But if the pitch to be "your own personal brand" had gone no further than the nation's convention halls and hotel ballrooms, just absorbed like so much cold coffee and muffin dribblings into the tattered carpet of the zeitgeist, I wouldn't be writing about it. The idea had legs, strong ones, and now you see whenever there's a public faux pas or a stumble from grace by some national figure, the natural question is: How will it affect his or her personal brand? Peters and Montoya were innovators, and I mean that sincerely. Some of the smartest and most deservedly successful people I know say the words "my brand" without irony. You can see the birth of the idea and its subsequent rise through mentions in print via Google Books.

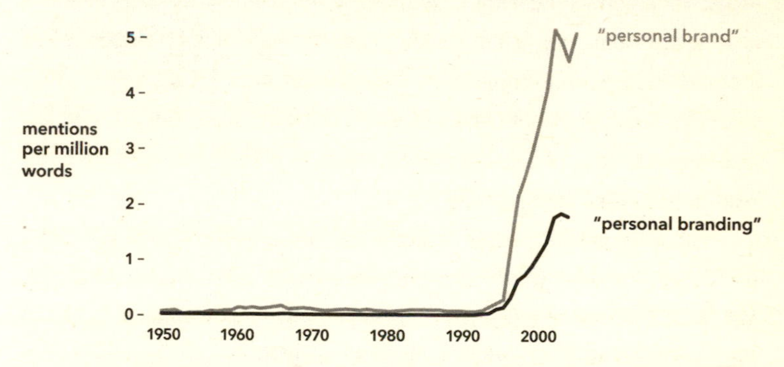

Of course, the principles of personal branding aren't new. Neither Montoya nor Peters* are all that different from Dale Carnegie, who rebranded himself from the plain "Dale Carnagey" by borrowing the golden surname of the steel magnate Andrew, and who, like these latter-day men, reduced character to bullet points and saw influence above all as the key to success. The goals of personal branding are the same you'd find in any empowerment seminar or in any prosperity gospel sermon from any decade. The end has always been wealth and power.

The new part is that "personal branding" asks you to accomplish these ends by treating yourself like a *product* rather than a human being. Peters again:

> Starting today you are a brand. You're every bit as much a brand as Nike, Coke, Pepsi, or the Body Shop. To start thinking like your own favorite brand manager, ask yourself the same ques-

* *His* mantra, by the way, is "distinct . . . or extinct."

tion the brand managers at Nike, Coke, Pepsi, or the Body Shop ask themselves: What is it that my product or service does that makes it different?

This is the core concept of personal branding, and like Christianity + the printing press or pro football + television, the idea has found in social media the perfect technology to go global. I won't rehash the ways sites like Facebook, Twitter, and Instagram give you the power to project yourself to the world. But I will point out that not long ago, only big companies, with big budgets, could get their message heard and beloved by strangers halfway around the globe. Now I can, and so can you, and so can everyone. The hardest part is getting anyone to listen.

The straightforward way is just to be entertaining, engaging, funny. But there's a reason comedians who can actually make people laugh are very rare. It's hard. An amateur who tries to build a following by being witty or provocative on Twitter is far more likely to end up the next Justine Sacco than the next Justin Halpern (@ShitMyDadSays), with his 3 million followers and a book deal. For every kid who tweets herself into college or into a cool job at the *New Yorker*—as people have done—there must be dozens who tweet themselves into the principal's office, or more likely, into a brick wall of embarrassed silence.

You can see something of what it takes to build a following using our text analysis algorithm. Here are the typical words for what I would call the "rank amateur" and "budding professional" follower levels:

most typical words for . . .

people with <100 followers	**people with 1,000+ followers**
#thehungergames	**partnering**
#upset	**#heyboo**
#worthit	**vamping**
#whyme	**optimizing**
roethlisberger	**sourcing**
workaholics	**marketer**

#wordsofwisdom	tweetup
#hurryup	visibility
#depressed	monetize
#wishmeluck	industry's
#getonmylevel	optimize
#studying	brownskin
#idiots	merchants
cincy	influencers
#collegeproblems	robust
#sunny	yeen
#notokay	guwop
#finalsweek	talmbout
#tebow	innovators
#silly	partnered
#impatient	bezos
#leavemealone	infographics
#holyshit	livest
#suckstosuck	strategist
pujols	entrepreneurial
#saveme	slideshare
#yeahbuddy	yass
pattys	amplify
#girlproblems	goodmorning
#killme	creatives

On the left you see the kinds of simple, fleeting concerns you'd expect from people on Twitter. On the right you see almost entirely management jargon: if you have a lot of followers, you are in fact much more likely to speak like a corporation. But some words on the right aren't typically professional: #heyboo, talmbout (a contraction of "talking about"), yeen ("you ain't"), yass ("your ass"), and a few others. Those are people using Twitter just like the folks on the left—to talk shit, complain, one-up—only they're doing it in wider circles, to thousands of followers. The users behind those words are black, and those terms' presence on the right side of the list is evidence of the different way African Americans tend to use the service. (I emphasize *tend* because no group is a monolith.) Observers call the phenomenon Black Twitter, described here by Farhad Manjoo in *Slate:*

> Black people—specifically, young black people—do seem to use Twitter differently from everyone else on the service. They form tighter clusters on the network—they follow one another more readily, they retweet each other more often, and more of their posts are @-replies—posts directed at other users. It's this behavior, intentional or not, that gives black people—and in particular, black teenagers—the means to dominate the conversation on Twitter.

By "dominate," he's referring to the fact that in Twitter's early years there was a lot of confusion from white users when hashtags like #uainthittinitright and #ifsantawasblack would make the service's Trending Topics list, alongside the latest deep thought from Ryan Seacrest or marketing gimmick from Old Spice (just as #heyboo might seem confusing alongside "monetize" above). Most users on Twitter follow institutions of one kind or another (celebrities, journalists, products) and those institutions don't follow them back. The mainstream culture of the service is organized around that one-to-many communication, organized, in fact, around the brand. But black users tend to focus on personal use and are highly reciprocal—hence high-follower counts and the enhanced ability to launch memes to the top of the charts.

Anyone hoping to build their brand on the service in the mainstream way—to become the one for the many—should realize that Twitter is very much the world of the One Percent. Its most precious resource, followers, is distributed far more unequally than wealth. In my sample, the top 1 percent of accounts has 72 percent of the followers. The top 0.1 percent has just over half. It is much, much harder to get to a million followers than it is to make a million dollars. There were 300,890 people who reported over $1 million in income to the IRS in 2011. Right now there are 2,643 Twitter accounts with 1 million followers, worldwide. Perhaps half are in the United States. Being an American with 1 million Twitter followers is roughly equivalent to being a billionaire.*

Of course, that assumes the followers are real. I bought some for one of my accounts to see how it works. On a site like TwitterWind, you can choose a number from a menu (I chose 1,000), pay up ($17), and a day or two later, and pretty much all at once, you get that many new, useless friends. The followers-for-hire do nothing at all but exist, and yet almost everyone with a really big Twitter following has probably bought some—especially the people for whom seeming popular is practically the whole job, like celebrities and politicians. When the Republican nomination was still up in the air, Newt Gingrich boasted, "I have six times as many Twitter followers as all the other candidates combined." The only catch was he'd paid for about 90 percent of them.† Mitt Romney (almost certainly) bought followers, too: for example, he gained 20,000 followers in a matter of minutes one day in July, which was about 200 times what he was getting immediately before and immediately after. Now, please note two important points: one, a person can buy followers for someone else, so this very well might've been some twenty-first-century Nixon working his ratfucking magic; it was certainly a good way to make Mitt look like a doofus. And, two, I'm sure Obama and many, many Democrats have bought followers for themselves. Craven attempts to game the system are a staple of both parties. They're just usually not as easy to catch as this:

* The 2014 Forbes Billionaires list has 1,645 members.

† One of Newt's former staffers told *Gawker:* "About 80 percent of those accounts are inactive or are dummy accounts created by various 'follow agencies,' another 10 percent are real people who are part of a network of folks who follow others back and are paying for followers themselves (Newt's profile just happens to be a part of these networks because he uses them, although he doesn't follow back), and the remaining 10 percent may, in fact, be real, sentient people who happen to like Newt Gingrich."

@MittRomney hourly activity, July 2012

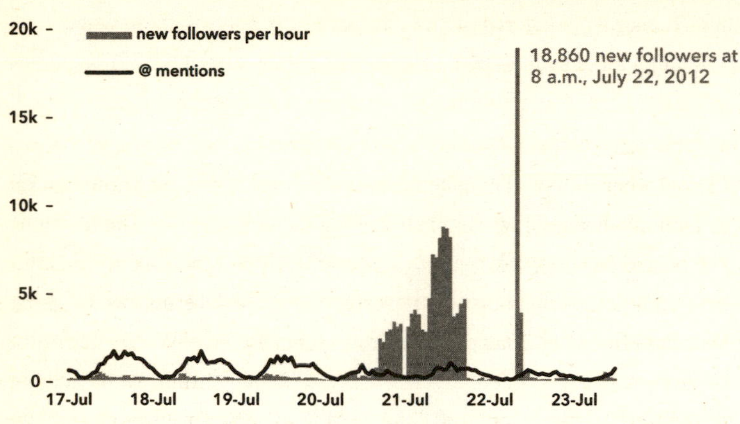

You can understand why these guys do it. The more popular someone *seems* to be, the more popular they become. It's as close as you can get to buying votes, at least until the Supreme Court makes that legal in 2018.

Everyday account holders are no less susceptible to the lure of easy friends, even if they don't have Barack's or Mitt's budget. Two of the five most common hashtags in my randomized Twitter data set (coming in at number one and number five, respectively) are #ff and #teamfollowback. The first stands for "Follow Fridays," which was an old-school tradition on Twitter—on Fridays you would tweet out people you like for your followers to follow. It's now just general (any-time) shorthand for "hey follow these accounts," and commonly blasted out by users just trying to drive numbers. The second, #teamfollowback, is the hashtag/handle for a Twitter account that basically does for free what politicians can afford to pay for. The idea is you follow TeamFollowBack, and the account's *other* followers will follow you. You then, in turn, follow *them* back, and everybody's numbers have risen. It's like the old idea of a "web ring," which in the days before Google was a way for websites to all link to one another and ensure traffic. It's also like the old idea of a full-on circle jerk. Here's TeamFollowBack's self-description:

> We will help you get followers that follow back! THE ORIGIONAL [*sic*] & THE BEST - Promote OUR hashtags #WILLFOLLOWBACK #TEAMFOLLOWBACK

So this is what the self-as-brand can lead to: chasing empty metrics. I know when I tweet, I'm as interested in who shares it, and how quickly, as I am in whatever I was originally trying to communicate. The few times I've posted to Facebook I've sat there and refreshed the page to catch the new comments, as though I'd never been on the Internet before. Jenna Wortham from the *Times* describes this mentality well: "We, the users, the producers, the consumers—all our manic energy, yearning to be noticed, recognized for an important contribution to the conversation—are the problem. It is fueled by our own increasing need for attention, validation, through likes, favorites, responses, interactions. It is a feedback loop that can't be closed, at least not for now." I can tell you from the inside: companies design their products to jam that loop open. OkCupid shows you little counts of your messages, your visitors, your possibilities. We know that those numbers keep our users interested, especially when they go up. Without little bits of excitement, a webpage or an app seems dead and people drift off. The broad term for this is "user engagement," how many people check in every week, every day, every hour. It's basically how fast they are running in the hamster wheel that's been set down for them there in cedar filings, and it's one of the most obsessed-over measures in the industry. Sites show you counts, totals, badges, because they know you'll come back to see them tick up. Then they can put your increased engagement on a slide to impress their investors.

That's the thing: it's one thing to reduce yourself to a number. When someone else reduces you, it feels ugly. Klout is one of the leading personal analytics firms; they look at all your social media accounts and, through a little proprietary black magic, give you an all-in measure of your online influence, 0 to 100. You'll remember per Montoya (and Carnegie): influence is what a personal brand is all about, and Klout helps you figure out how you're doing. Right now, my Klout score is a fairly pathetic 34. TeamFollowBack comes in at 60, which makes me want to either laugh or cry. On the one hand, these people have gotten the equivalent of a D– grade on their only reason to exist. On the other, they have a higher score than anyone I know.

In 2012, Salesforce.com, the cloud-computing behemoth, posted a job opening that listed a Klout score of at least 35 as a "desired skill." It wasn't positioned as a requirement, but they put it up there along with the allow-us-to-state-the-obvious attributes like "ability to work . . . as a part of a team," so it was presumably a core part of the job. Salesforce's business specialty is quantification—they help companies market through data.* So it's not that surprising that they would approach hiring in the same quantified way. But even though numbers like credit scores have been an odious part of the HR process for some time, seeing a Klout score on a job listing got a lot of people upset.

BetaBeat's article "Want to Work at Salesforce? Better Have a Klout Score of 35 or Higher" got the general reaction just right with their one-word subhead: "Ugh." However, the real concern—that we're all going to be reduced to numbers, and soon—deserves a longer discussion. Salesforce was, and is, a trendsetter—certainly in the world of online marketing. They were *Forbes*'s "Most Innovative Company in America" the same year they put up that post. They hire hundreds of people a year, and, even more to the point, when award-winning innovators do something new, other companies copy it. If Salesforce is asking for Klout scores, then everyone will soon be asking for Klout scores. People don't want to be reduced to a two-digit number, concocted by a company that even in the vaporous world of social media startups seems kind of bullshitty.

But given that Klout uses many of the same reductive tools that I myself have employed to gather data, where does that leave you and me and the book we've both spent all this time with? Well, the short answer is: right there with Klout and Salesforce. Reduction is inescapable. Algo-

* As an analytics bona fide, they even own data.com.

rithms are crude. Computers are machines. Data science is trying to make digital sense of an analog world. It's a by-product of the basic physical nature of the microchip: a chip is just a sequence of tiny gates. Not in the way that the Internet is a "series of tubes" but in actuality. The gates open and close to let electrons through, and when one of these gates wants to know what state to be in, it's all or nothing—like any door, a circuit is open or it isn't; there are no shades of maybe. From that microscopic reality an absolutism propagates up through the whole enterprise, until at the highest level you have the definitions, data types, and classes essential to programming languages like C and JavaScript.

Thus, information is reduced by necessity. But fundamentally the objections to the Klout-score requirement were about the *people* being reduced to digits, not just their information. And here's where *Dataclysm* diverges from Salesforce's job post, and indeed Klout's whole business model.

As many numbers as there are here, they're not meant to stand in for any one person. A single number never could. It's a truth summed up by the apocryphal story that Einstein flunked math in high school. He didn't. But he could've, and if he had, who cares? If he got a 35 in Algebra II, so what? Is he suddenly not smart? No number, no test, no single measurement—not IQ, not height, and certainly not a Klout score or friend count or reply percentage on OkCupid—is a whole person, which is exactly why, beyond illustration, individual users don't appear in this book. But by aggregating a bunch of these small and inadequate parts of us together, we get something big. The law of large numbers is an idea we've brushed past a few times, but I want to lay it out explicitly: the full truth of data is only revealed over a large sample. Imagine a mysterious die—you can't count the sides but you can roll it and see what comes up. Roll once and you could get any number, you learn nothing. Roll it a bunch of times, you get the distribution, you get the average—and that defines the die right there. You know the shape only through aggregation.

What's more, reduction and repetition are fundamental to the long history of science, not just data science and not just computer science, but capital-S Science, the ageless human enterprise. Experiments are built upon reducing a process to a single, manageable facet. The scientific method needs a control, and you can't get it without cutting complexity to the bald core and saying this, *this,* is what matters. Only once you've simplified the question can you test it over and over again. Whether at a lab bench or a laptop, most of the knowledge we possess was acquired like this, by reduction.

So here, we've boiled humanity down to numbers rather than, say, anecdotes. In my mind—and this takes nothing away from Malcolm Gladwell—I see this book as the opposite of outliers. Instead of the strays from the far reaches of the data—the one-offs, the exceptions, the singletons, the Einsteins for whom you need the whole story to get it right—I'm pulling from the undifferentiated whole. We focus on the dense clusters, the centers of mass, the data duplicated over and over by the repetition and commonality of our human experience. It's science as pointillism. Those dots may be one fractional part of you, but the whole is us.

Aggregation and reduction also allow us to deal in broad trends, the smooth flow of which might not have the peaks and troughs of the usual hero narratives but which are all the more applicable for it. The fact that Paul McCartney and John Lennon practiced rock music for 10,000 hours and then became the Beatles does say something about the value of rehearsal and persistence, but that number itself means nothing. I myself have put in that kind of time playing guitar, as have many others whose music you'll never hear. Whatever it was that allowed Lennon and McCartney to turn practice into genius, it's unique to them. On the other hand, every number in this book has many hundreds, often many thousands, of people behind it, none of them famous. Here's the kernel of it: the phrase "one in a million" is at the core of so many wonderful works of art. It means a person so special, so talented, so *something* that they're practically unique, and that very rareness makes them significant. But in mathematics, and so with data, and so here in this book, the phrase means just the opposite: 1/1,000,000 is a rounding error.

But if simplifying is what it takes to understand large data sets, I do worry about a different kind of reductionism: people becoming not a number exactly, but a dehumanized userid fed into the grind of a marketing algorithm; grist for someone else's brand. Data takes too much of the guesswork out of the sell. It's a rare urban legend that turns out to be true, but Target, by analyzing a customer's purchases, really did know she was pregnant before she'd told anyone. The hitch was that she was a teenager, and they'd started sending maternity ads to her father's house.

In some ways, that kind of corporate intrusion is better than brands actually trying to "relate." Last summer, a Jell-O marketing campaign co-opted (tagjacked?) the hashtag #fml, which is Internet shorthand for "fuck my life." Their social media people began responding to tweets that contained the tag with an unsolicited offer to "fun" the person's life instead, with coupons. Thus people in extremis received jaunty offers from a gelatin, as in this exchange:

> Pyrrhus Nelson @suhrryp
>
> Seeing my bank account disappear at the dr office #fml
>
> JELL-O @JELLO
>
> @suhrryp Fun My Life? Of course we will. In fact, we'd be happy to. prmtns.co/dkTq Exp. 48hrs

This kind of unwanted intercession is all too easy on social media because everything is so quantified. The hashtags jump right to the brand manager's screen; he dives in with the discounts. At least the same technology that allows them into our lives allows us to fight back. A few years ago, McDonald's sent out a couple tweets, feel-good stories about their suppliers, with the tag #McDStories, and they got #fml'd in reverse. This is just one of many responses:

> MUZZAFUZZA @Muzzafuzza
>
> I haven't been to McDonalds in years, because I'd rather eat my own diarrhea. #McDStories

McDonald's had paid to promote the hashtag and pulled the campaign after only a couple hours when it quickly spiraled out of their control. A week later, the repurposed #McDStories was still going strong. Their social media strategists should've known what to expect: a few months before, Wendy's had tried to push #HeresTheBeef, and their catchphrase was ripped completely free of the intended context. People used it to complain about anything they didn't like (had a beef with), ignoring the brand:

> Remi Mitchison @RemiBee
> #HeresTheBeef when a chick see another chick doin better and has more than she does . . . so she wanna stunt and #GetThatAssBeatUp

> Jeremy Baumhower @jeremytheproduc
> #HeresTheBeef The drugs companies have already cured HIV and cancer, however it is far more profitable to keep people barely alive on drugs

More recently, Mountain Dew ran a "Dub the Dew" contest, trying to ride the "crowdsourcing" wave to a cool new soda name and thinking maybe, if everything went just right and the metrics showed enough traction to get buy-in from the right influencers, they'd earn some brand ambassadors in the blogosphere. Reddit and 4chan got ahold of it, and "Hitler did nothing wrong" led the voting for a while, until at the last minute "Diabeetus" swooped in and the people's voice was heard: *Dub yourself, motherfucker.*

The Internet can be a deranged place, but it's that potential for the unexpected, even the insane, that so often redeems it. I can't imagine anything worse for You! The Brand! than upvoting Hitler. Plus, what a waste of time, because obviously Mountain Dew isn't going to print a single unflattering word in the style of its precious and distinctive marks. I find comfort in the silliness, in the frivolity, even in the stupidity. Trolling a soda is something no formula would ever recommend. It's no industry best practice. And it's evidence that as much as corporatism might invade our newsfeeds, our photostreams, our walls, and even, as some would hope, our very souls, a small part of us is still beyond reach. That's what I always want to remember: it's not numbers that will deny us our humanity; it's the calculated decision to stop being human.

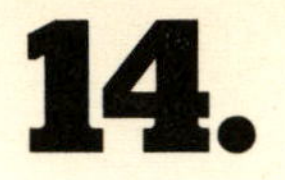

Breadcrumbs

Facebook released the Like button in 2009 and it changed the way people shared content. The idea wasn't new—once-popular, now marginal, sites like digg.com and del.icio.us had been letting people "like" articles for years before that. But at these companies, the content was the star. Facebook laid curation over an already robust social network and, for the content creators, made it simple for anyone to attach that iconic little thumbs-up to their work. They created a new universal microcurrency—I might not *pay* you for your writing, music, or whatever, but I'll give you a fillip of approval and share what you've done with my friends. As of May 2013, Facebook was recording 4.5 billion likes *a day* and in September of that year reported that 1.13 trillion had been submitted all-time.

Those students from MIT developed their gaydar the same year likes launched. Their algorithm was pretty good at guessing a man's sexuality, but it also worked in a fairly obvious way: it's surely no big secret that gay men are more likely to have gay male friends. The gaydar innovation was to use macro-level data to do something people had been doing in small ways all along. Since then, the power of predictive software has advanced rapidly; these types of programs only get smarter and faster as more data becomes available. By 2012, a group from the UK had discovered that from a person's likes alone they could figure out the following, with these degrees of accuracy:

whether someone is . . .	
Caucasian or African American	95%
a man or a woman	93%
gay or straight	88%
Democrat or Republican	85%
lesbian or straight	75%
a drug user	65%
the child of parents who got divorced before he or she turned 21	60%

Again, this is not from looking at status updates or comments or shares or anything that the users typed. Just their likes. You know the science is headed to undiscovered country when someone can hear your parents fighting in the click-click-click of a mouse. A person's "like" pattern even makes a decent proxy for intelligence—this model could reliably predict someone's score on a standard (separately administered) IQ test, without the person answering a single direct question.

This stuff was computed from three years of data collected from people who joined Facebook after decades of being on Earth without it. What will be possible when someone's been using these services since she was a child? That's the darker side of the longitudinal data that I'm otherwise so excited about. Tests like Myers-Briggs and Stanford-Binet have long been used by employers, schools, the military. You sit down, do your best, and they sort you. For the most part, you've opted in. But it's increasingly the case that you're taking these tests just by living your life. And the results are there for anyone to read and judge. It's one thing to see that someone's Klout score is 51 or whatever in advance of a job interview. It's another to know his IQ.

If employers begin to use algorithms to infer how intelligent you are or whether you use drugs, then your only choice will be to game the system—or, to borrow the wording from the previous chapter, "manage your brand." To beat the machine, you must act like a machine, which means you've lost to the machine. And that's all assuming you can guess at what you're supposed to do in the first place. Apparently, one of the strongest correlates to intelligence in the research was liking "curly fries." Who could reverse-engineer that?

But while Facebook does know a lot about you, it's more like a "work friend"—for all the time you spend together, there are clear limits to your relationship. Facebook only knows what you do on Facebook. There are many places with much deeper reach. If you have an iPhone, Apple could have your address book, your calendar, your photos, your texts, all the music you listen to, all the places you go—and even how many steps it took to get there, since phones have a little gyroscope in them. Don't have an iPhone? Then replace "Apple" with Google or Samsung or Verizon.

Wear a FuelBand? Nike knows how well you sleep. An Xbox One? Microsoft knows your heart rate.* A credit card? Buy something at a retailer, and your PII (personally identifiable information) attaches the UPC to your Guest ID in the CRM (customer relations management) software, which then starts working on what you'll want next.

This is just a sliver of the corporate data state, the full description of which could take pages. For the government picture, a sliver is all I have, because that's all we've been able to see of it. We do know that the UK has 5.9 million security cameras, one for every eleven citizens. In Manhattan, just below Fourteenth Street, there are 4,176. Satellites and drones complete the picture beyond the asphalt. Though there's no telling what each one sees, it's safe to say: if the government is interested in your whereabouts, one sees you. And besides, as Edward Snowden revealed, much of what they can't put a lens on they can monitor at leisure from the screen of an NSANet terminal, location undisclosed.

Because so much happens with so little public notice, the lay understanding of data is inevitably many steps behind the reality. I have to say, just pausing to write this book, I'm sure I've lost ground. Analytics has in many ways surpassed the information itself as the real lever to pry. Cookies in your web browser and guys hacking for credit card numbers get most of the press and are certainly the most acutely annoying of the data collectors. But they've also taken hold of a small fraction of your life, and for that small piece they had to put in all kinds of work. No matter how crafty the JavaScript, they're villains in the silent-film vein, all mustachios and top hats. Or, a more contemporary reference: they're like so many pasty Dr. Evils—underworld relics holding the world hostage for *one . . . million . . . dollars* . . . while the

---

* From *Nature*'s discussion of the console: "It is fitted with a camera that can monitor the heart rate of people sitting in the same room. The sensor is primarily designed for exercise games, allowing players to monitor heart changes during physical activity, but, in principle, the same type of system could monitor and pass on details of physiological responses to TV advertisements, horror movies or even . . . political broadcasts."

billions fly by to the real masterminds, like Acxiom. These corporate data marketers, with reach into bank and credit card records, retail histories, and government filings like tax records, know stuff about human behavior that no academic researcher, fishing for patterns on some website, ever could.* Meanwhile, the resources and expertise the national security apparatus brings to bear makes enterprise-level data-mining software look like Minesweeper.

This data, despite the "mining" metaphor, isn't a naturally occurring resource; it comes from somewhere—and that somewhere is you. The companies and the government are collecting disparate pieces of your private life and trying to fashion them back into an image they can master. The more privacy you lose, the more effective they are. The fundamental question in any discussion of privacy is the trade-off—what *you* get for losing it. We make calculated trades all the time. Public figures sell their personal lives to advance their careers. Anyone who's booked a hostel in Europe or bought a train ticket in India has had to decide if the private room is worth the extra money. And not to confuse the issue here, but many people, men and women, trade on privacy when they walk out the door in the evening, giving it away, via a hemline or a snug fit, for attention. So the exchange isn't new. But our trading partners, and their terms, are. On the corporate side, the upshot of our data (the benefit to us) isn't all that interesting unless you're an economist. In theory, your data means ads are better targeted, which means less marketing spend is wasted, which means lower prices. At the very least, the data they sell means you get to use genuinely useful services like Facebook and Google without paying money for them. What we get in return for the government's intrusion is less straightforward.

Does surveillance make us more safe? Is the security apparatus a

* From Acxiom's website: "[We give] our clients the power to successfully manage audiences, personalize customer experiences and create profitable customer relationships." An interesting paradox: whenever you see the word "personalize," you know things have gotten very impersonal.

blanket? Well, there haven't been any terror attacks on American civilians since 2001—at least, not ones by the syndicates. That's not meaningless, certainly not to a New Yorker. But an argument from absence isn't very strong, and at least until we're allowed to know the threats that were thwarted as opposed to those never planned, it's hard to trust what we're told. Like so much Texas dust, its memory has almost drifted away, but the color-coded "Threat Level" that was such a part of the discussion in the years after 9/11 always felt to me like an elaborate advertisement for Halliburton. It's hard to believe in information coming to you on a "need to know basis" from an entity that doesn't think you need to know anything. The concern becomes less about what they're saying than why. In any event, I have no idea how many, if any, crimes the big glean at the NSA has prevented. We're told it works, just not when, where, or how.

Quixotically, for those crimes total surveillance *didn't* prevent, it has certainly proved useful in solving. All those security cameras cracked the case after the Boston Marathon bombing, as they did after the London subway bombings in 2005.* Especially for asynchronous crimes, you need total data to return to, because the criminals commit their acts long before any victims fall. In these investigations, the power of the intelligence becomes part of the media story—this is the surveillance state's time to shine. The data has a defined purpose, and no one debates the privacy/protection balance while there is blood on the ground. But in between the times of "United We Stand" a lot of what we learn about what the government knows comes from whistle-blowers like Snowden.

The NSA is the government's signals intelligence arm, and here the signal they're looking for is in our data. I have some personal familiarity with the organization. As I've said, I studied math. I did so at Harvard. My

* After Boston, Reddit and 4chan tried vigorously (meaning there was lots of typing) to track down the bombers and eventually "pinned" it on an innocent man. For all the lip service the cloud and crowd get, hardware solved the crime.

bachelor's degree looks just like my classmates', but there were unofficially two tracks in the department. One, mine, was for the kids who liked math and were pretty good at it. The other was for the transcendent savants. There was a difficult first-year course called Math 25, which I wasn't good enough for, and from which the ultra-elite were drawn into a superclass called Math 55 by special invitation from the department. The hardest courses I ever took were often entirely skipped by these real mathematicians. The teaching assistants in my high-level courses, the people who handled a lot of the actual instruction and all of the grading, were not only often younger than me (one was sixteen) but were already deep into the graduate-level curriculum, as teenagers. I remember being very excited about (and challenged by) Real Analysis, which was a class that many of my peers—as if that's the right word—would've found boring as ninth-graders. Whenever I hear the letters "NSA," I think back to those days, because they recruited from that second track.

I point this out because, to many people, government workers have an indifferent reputation—bureaucrats, functionaries, whatever. And certainly the average person working in data analytics in the private sector is as likely to be competent as not. But the people spying on us are extremely, extremely smart. We can hope that they, like Feynman and Einstein before them, are able to temper their work with a farsighted humanity, but we can *know*, for sure, that, like Feynman and Einstein before them, what they're working on is inhumanly powerful.

Insofar as algorithms are fed by data, Mr. Snowden has revealed that the NSA's are fatted on superfood. Or rather . . . all the food. They gather phone calls, e-mail, text messages, pictures, basically everything that travels by electric current. It's clear that it's not a passive operation—according to one leaked document, the stated, top-level purpose is to "master the Internet." The project's brazenness is one of the most phenomenal things about it. Among the first documents published (jointly by the *Guardian* and the *Washington Post*) was a PowerPoint presentation about a program called PRISM. The slides don't beat around the bush:

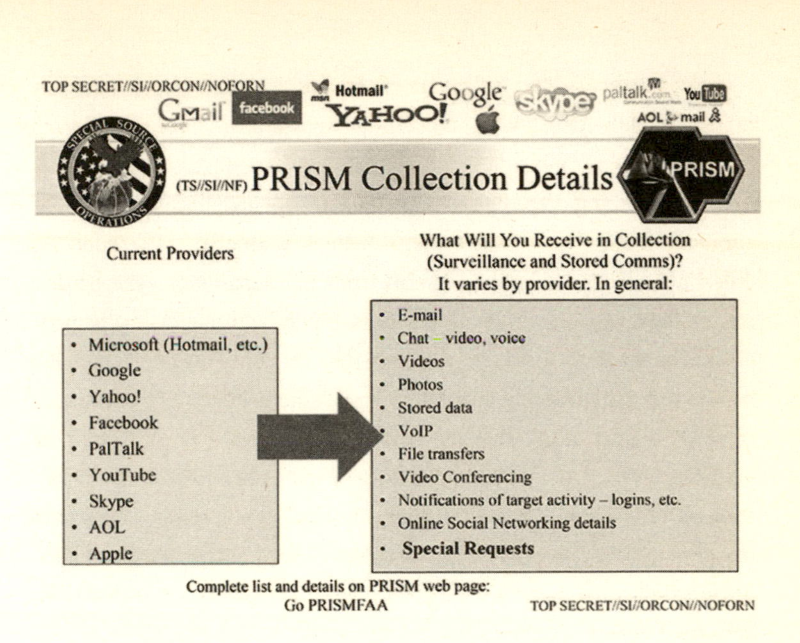

It should've been called Operation Yoink! On the one hand, life on Earth only gets worse when anyone wearing a sidearm starts thinking about our Facebook accounts. On the other, it's hard to be afraid of people using the Draw tool in a Microsoft product.

No one sees the PRISM data for an individual without a court order, at least in theory, because the program is so invasive. Other snooping is mostly focused on metadata—the incidentals of communication. Here's the government's own Privacy and Civil Liberties Oversight Board describing one part of another project:

> For each of the millions of telephone numbers covered by the NSA's Section 215 program, the agency obtains a record of all incoming and outgoing calls, the duration of those calls, and the precise time of day when they occurred. When the agency targets a telephone number for analysis, the same information [is obtained] for every telephone number with which the original

number has had contact, and every telephone number in contact with any of those numbers.

It must be said that none of this entails the actual *content* of anyone's communication. In that respect, it's not much different from the data we've looked at in this book. We let patterns stand in for any single person's life, just like these guys do. At the NSA, again according to them, if your web of calls fits the profile of a "threat," only then do they start paying real attention. But metadata isn't necessarily less invasive for being indirect.

People leave some amazing breadcrumbs for anyone interested in following them. You've seen plenty already—200 pages' worth. Even so, there are just as many trails we haven't followed. For example, a little text file called the Exif is attached to all images taken with a digital camera, from high-end SLRs to your iPhone. The file encodes not only when the picture was taken but miscellany like the *f*-stop and shutter speed for the photo and, often, the latitude and longitude of where it was taken. Exif is how programs like iPhoto can effortlessly sort your pictures into "moments" and place little pins all over the map to show you where you've been. There are other things the Exif can tell you, though. Take the profile photos on OkCupid. The better-looking a photo is, the better chance it has of being outdated. That is, people find that one "great picture" and just lock it in forever. We know this because of the Exif, which tells us when the picture was taken. This kind of data tagalong is common. GPS coordinates ride shotgun over the network whenever you open your favorite app. Almost every web page you've ever loaded has dozens of one-pixel images (just a single transparent dot) buried in the margins that, by being loaded alongside the "real" page, register your visit; the pixels can't tell what you're doing, just when and where you've gone. This simple stuff, just *whens* and *wheres* can give a company a good guess at your whole demographic profile.

What about the people who don't want to share like this? The people who would rather shop and preen alone? I myself know the value of privacy. That's part of the reason I'm not a big social-media user, frankly.

I have never posted a picture of my daughter on the Internet. I started using Instagram in earlyish 2011 when the service wasn't big yet, and I used it as just a photo gallery app because I liked the filters. I thought it was like Hipstamatic, not really social—I know this makes me sound like a grandfather. When my wife realized what her fuddy-duddy husband was doing, she pointed out that I could connect my account to other people's accounts, which I did, because hey, look: a button to click. But once it wasn't just me on my own with my pictures, it lost all appeal.

This kind of reticence is unusual. For all the hand-wringing, it's hard to argue that most users are anything but blasé about privacy. Whenever Facebook updates its Terms of Service to extend their reach deeper into our data, we rage in circles for a day, then are on the site the next, like so many provoked bees who, finding no one to sting, have nowhere to go but back to the hive. Because tech loves to push boundaries and the boundaries keep giving, software has gotten almost aggressively invasive. There are weight-loss apps. Heart-rate apps. Rate-my-outfit apps—submit your ensemble to the crowd for fashion advice. Women are using apps to predict and manage their menstrual cycle: "The market is flooded with them," as Jenna Wortham writes, before adding, "nearly every woman I know uses one." You let the app know when your period starts, and it'll alert you when you're at peak fertility, to avoid or embrace as you wish. Of course, self-reported data not being quite invasive enough, there's a startup that says it can *infer* when a woman is having her period from her link history. Any of these menstruation apps—at least if they have a competent data scientist behind them—will of course also know when a user is pregnant, overexercising, getting older, or having unprotected sex, since when you're late, you'll check the thing unusually often.

But despite some, even many, people's cavalier attitude toward privacy, I didn't want to put anyone's identity at risk in making this book. As I've said, all the analysis was done anonymously and in aggregate, and I handled the raw source material with care. There was no personally identifiable information (PII) in any of my data. In the discussion of users' words—their profile text, tweets, status updates, and the like—those words were public. Where I had user-by-user records, the userids were encrypted. And in any analysis the scope of the data was limited to only the essential variables, so nothing could be tied back to any individual.

I never wanted to connect the data back to individuals, of course. My goal was to connect it back to everyone. That's the value I see in the data and therefore in the privacy lost in its existence: what we can learn. Jaron Lanier, author of *Who Owns the Future?* and a computer scientist currently working at Microsoft Research, wrote in *Scientific American* that "a stupendous amount of information about our private lives is being stored, analyzed and acted on in advance of a demonstrated valid use for it." He's unquestionably right about the "tremendous amount," but I take issue with his final clause. How does anything ever become useful if it can't be "acted on in advance of a demonstrated valid use"? The whole idea of research science is predicated on exploration. Iron ore was once just another rock until someone started to experiment with it. Mold on bread spent millennia just making people sick until Alexander Fleming discovered it also made penicillin.

Already data science is generating deep findings that don't just describe, but change, how people live. I've already mentioned Google Flu; launched in 2008, it now tracks nascent epidemics in more than twenty-five countries. It's not a perfect tool, but it's a start. Combined data is even being used to prevent disease, not just minimize it. As the *New York Times* reported last year: "Using data drawn from queries entered into Google, Microsoft and Yahoo search engines, scientists at Microsoft, Stanford and Columbia University have for the first time been able to detect evidence of unreported prescription drug side effects before they were found by the Food and Drug Administration's warning system." The researchers determined that paroxetine and pravastatin were causing hyperglycemia in patients. Here, the payoff for living a little less privately is to live a little more healthily.

Every day, it seems, brings word of some new advance. Today, I found out that a site called geni.com is well on the way to creating a crowdsourced family tree for all mankind. If it works, the company will have made, essentially, a social network for our genetic material. The week before, two political scientists debunked the received wisdom that Republicans owe their House majority to district gerrymandering. The authors had modeled every possible election over every possible configuration of the United States and concluded, with the computer playing Candide, that our divided world is the best we can hope for. The political geography of the country, not the actual maps, creates the gridlock.

This is just the beginning. Data has a long head start—Facebook was collecting 500 terabytes of information every day way back in 2012—but the analysis is starting to catch up. Data journalism was brought to the mainstream by Nate Silver, but it's become a staple of reporting: we quantify to understand. The *Times*, the *Washington Post*, the *Guardian* have all built impressive analytic and visualization teams and continue to devote resources to publishing the data of our lives, even in the constrained financial climate for reporters and their work.

On the flush corporate side, Google, mentioned many times in these pages, leads the way in turning data to the public good. There's Flu and the work of Stephens-Davidowitz, but also a raft of even more ambitious, if less publicized, projects, such as Constitute—a data-based approach to constitution design. The citizens of most countries are usually only concerned with one constitution—their own—but Google has assembled all nine hundred such documents drafted since 1787. Combined and quantified, they give emerging nations—five new constitutions are written every year—a better chance at a durable government because they can see what's worked and what hasn't in the past. Here, data unlocks a better future because, as Constitute's website points out: in a constitution, "even a single comma can make a huge difference."

As we've seen, Facebook's data team has begun to publish research of broad value from their immense store of human action and reaction. Seizing on that Newtonian interplay, Alex Pentland at MIT calls the emerging science "social physics." He and his team have begun moving social data to the physical world. Working with local government, communications providers, and citizens, they've datafied an entire city. The residents of

Trento, Italy, can now tackle, with hard numbers, what for the rest of us are workaday unanswerables: "How do other families spend their money? How much do they get out and socialize? Which preschools or doctors do people stay with for the longest time?"

Perhaps this is the future we have to look forward to. I've tried to explain what we've already learned by combining the best of the work that's out there with my own original research. In so doing, more than stretching out my arms to say *This is the pinnacle,* I mean to communicate the power of what's to come. Watson and Crick unlocked the secret of DNA in 1953, and six decades later scientists are still decoding the human genome. The science of our shared humanity—the search for the full expression of the genes we'll soon have fully mapped—is years from anything so lofty.

As far as balancing the potential good with the bad, I wish I could propose a way forward. But to be honest I don't see a simple solution. It might be that I'm too close. I share Lanier's belief that regulation won't work. Not that someone won't try that route. The new laws will be drafted with all the right spirit, I'm sure, but their letter will be outdated before the ink is dry. And being on the data collectors' side myself, I've seen first-hand that you can give people all the privacy controls in the world, but most people won't use them. OkCupid asks women: *Have you ever had an abortion?*—it's the 3,686th match question; I told you they truly cover everything. Right beneath the question, there's a checkbox to keep your answer private. Of the people who answer in the affirmative, fewer than half check the box.

So most people won't use the tools you give them, but maybe "most people" is the wrong goal here. For one thing, providing ways to delete, or even repossess, data is the right thing to do, no matter how few users take you up on it. For another, it's possible that privacy has changed, and left the people writing about it behind. Lanier and I are old men by Internet standards, and it's not just in armies that "generals always fight the last war." My expectations of what is correct and permissible might be wrong. Cultures and generations define privacy differently.

People aren't even that upset about the NSA, as gross as their overreach is. There have been many "Million" marches on Washington. Million Man, Million Mom, and so on. Recently, the hacker collective Anonymous called for a Million Mask March to protest, among other things, the PRISM program and government mass surveillance. The *Washington Post* captures the shortfall of public interest in just the first word of their coverage: "Hundreds of protesters . . ."

In his *Scientific American* piece, Lanier proposes that we be compensated for our personal data and let market forces rebalance the privacy/value equation. He proposes that data collectors issue micropayments to users whenever their data is sold. But that expense, like a tax, either will be passed directly back to the consumer or will bring on a race to the bottom, where websites have to find margin wherever they can get it, the way commercial airlines do now. Either way, there's no net value in it for us. And that's not to mention the impracticality of making it happen.

Pentland's approach is much more feasible: he calls it his "New Deal on Data." Ironically enough, it harkens back to Old English Common Law for its principles. He believes that, as with any other thing you own, you should have the fundamental rights of possession, use, and disposal for your data. What that means is you should be able to remove your data from a website (or other repository) whenever you feel like it's being misused. You should also be allowed to "take it with you," in theory for resale, should a market for that develop. That simple mechanism—the Delete button, with the option to copy/paste—is not only more feasible but also more fair than any enforced compensation.

In fact, I would argue that people are *already* compensated for their data: they get to use services like Facebook and Google—connect with old friends, find what they're looking for—for free. As I've said, I give these services little of myself; but I get less out of them too. People have to decide their own trade-off there. Soon, though, there might be only one decision to make: am I going to use these services at all? The analytics are becoming so powerful that it may not matter what you try to hold back. From only the barest information, algorithms are already able to extrapolate or infer much about a person; that's after only a few years of data to work on. Soon the half measures provided by menu options as you "manage your privacy settings" will give no protection at all, because the rest of your world won't be so withholding. Companies and the government will find you through the graph. This whole debate could soon be an anachronism.

In any event, when I talked about the data as a flood way, way back, I perhaps didn't emphasize it enough: the waters are still churning. Only when they start to calm can people really know the level and make good the surfeit. I am eager to do so. In the meantime, the people who store, analyze, and act on data have a responsibility to continue to prove the value of their work—and to reveal exactly what it is they're doing. Or else, for all my quibbling, Lanier is right: we shouldn't be doing it.

Technology is our new mythos. There's magic in some of it, undeniably. But even grander than the substance is the image. Tech gods. Titans. Colossi astride the whole Earth, because, you know, Rhodes just isn't cool anymore. This is how the industry is often cast to the public, and sadly it's how it often thinks of itself. But though there are surely monsters, there are no gods. We would all do well to remember this. All are flawed, human, and mortal, and we all walk under the same dark sky. We brought on the flood—will it drown us or lift us up? My hope for myself, and for the others like me, is to make something good and real and human out of the data. And while we do, whenever the technology and the devices and the algorithms seem just too epic, we must all recall Tennyson's aging Ulysses and resolve to search for *our* truth in a slightly different way. To strive, to seek, to find, but then, always, to yield.

Coda

Designing the charts and tables in this book, I relied on the work of the statistician and artist Edward R. Tufte. More than relied on, I tried to copy it. His books occupy that smallest of intersections: coffee-table beautiful and textbook clear, and inside he lays out principles of information design drawn from the all-time famous examples of data as storytelling. Charles Minard's plot of Napoleon's Russian undoing. An unnamed abolitionist's *Description of a Slave Ship*, showing the human cargo packed in inhuman closeness, an image that is still the iconic shorthand for the horrors of the Middle Passage. Dr. John Snow's plot of a cholera outbreak in 1854 pinpointed the source of the disease for the first time. Tufte pulls lessons from these and makes them useful in a modern context, asking the data designer to maximize the data-to-ink ratio. Give every chart a clear story to tell. Use white as dimension, not dead space. I've tried my best.

Among the many maps and charts and tables in Tufte's books, there's a two-page examination of the Vietnam Memorial, not as stonework or as history, but as an artifact of data design. I wish I could reprint the full discussion here, but the kernel is this.

> From a distance the entire collection of names of 58,000 dead soldiers arrayed on the black granite yields a visual measure of what 58,000 means, as the letters of each name blur into a gray shape, cumulating to the final toll.

To find meaning in that gray blur is what every data scientist hopes for, and I've sought that distance and that blur repeatedly in these pages, drawing from the biggest data sets, looking at the widest stories, all to better my chances at truth.

The memorial was digitized in 2008. Every square inch was photographed and collated with military records, and the online version allows visitors to attach photos and text to each name. The web archive confronts the visitor with an empty box, demanding, "Search the Wall." After a pause, I started to type my dad's name, because when I think of Vietnam I think of him almost as a reflex. But then I remembered, gratefully, David Patton Rudder isn't on this list. So I entered *someone's* name, just a guess—"John" of course and then because Smith seemed too bland and Doe too hokey, "Wilson." The page churned for a half second, and at the top I saw:

Lorne John Wilson	
Tour Start Date	1969-03-17
Tour End Date	1969-03-28
Death Date	1969-03-28
Age	20

Two pictures had been added to his entry, one his portrait in dress blues, the other a snapshot, perhaps taken one of those eleven days PFC Wilson was in-country and alive. It shows four young men around a jeep, one's standing in the back; they're just talking in the afternoon. Grainy and undersaturated, but for the fatigues it could've come from Instagram. Whoever uploaded it had held on to the picture, and his friends, for decades.

A web page can't replace granite. It can't replace friendship or love or family, either. But what it can do—as a conduit for our shared experience—is help us understand ourselves and our lives. The era of data is here; we are now recorded. That, like all change, is frightening, but between the gunmetal gray of the government and the hot pink of product offers we just can't refuse, there is an open and ungarish way. To use data to know yet not manipulate, to explore but not to pry, to protect but not to smother, to see yet never expose, and, above all, to repay that priceless gift we bequeath to the world when we share our lives so that other lives might be better—and to fulfill for everyone that oldest of human hopes, from Gilgamesh to Ramses to today: that our names be remembered, not only in stone but as part of memory itself.

A Note on the Data

Numbers are tricky. Even without context, they give the appearance of fact, and their specificity forbids argument: *20,679 Physicians say "LUCKIES are less irritating."* What else is there to know about smoking, right? The illusion is even stronger when the numbers are dressed up as statistics. I won't rehash the old wisdom there. But behind every number there's a person making decisions: what to analyze, what to exclude, what frame to set around whatever pictures the numbers paint. To make a statement, even to just make a simple graph, is to make choices, and in those choices human imperfection inevitably comes through. As far as I know, I've made no motivated decision that has bent the outcome of my work—the data of people acting out their lives is interesting enough without me needing to lead it one way or another. But I have made choices, and those choices have affected the book. I'd like to walk you through a few of them.

My first choice was probably my most difficult: the decision to focus on male-female relationships when I talk about attraction and sex. Space, of course, was a factor—to include same-sex relationships would've meant repeating each graph or table in triplicate. But more than that was the discovery that same-sex relationships aren't exceptional—they follow all the same trends. Gay men, for example, prefer younger partners just like straight men do. For issues that have to do with sex only indirectly, such as ratings from one race to another, gays and straights also show similar patterns. Male-female relationships allowed for the least repetition and widest resonance per unit of space, so I made the choice to focus on them.

My second decision, to leave out statistical esoterica, was made with much less regret. I don't mention confidence intervals, sample sizes, p values, and similar devices in *Dataclysm* because the book is above all a popularization of data and data science. Mathematical wonkiness wasn't what I wanted to get across. But like the spars and crossbeams of a house, the rigor is no less present for being unseen. Many of the findings in the book are drawn from academic, peer-reviewed sources. I applied the same standards to the research I did myself, including a version of peer-review: much of the OkCupid analysis was performed first by me and then verified independently by an employee of the company. Also, I separated the analysis from the selection and organization of the data to make sure the former didn't motivate the latter. One person would extract the information, another would try to figure out what it meant.

Sometimes, I present a trend and attribute a cause to it. Often that cause is my best guess, given my understanding of all the forces in play. To interpret results—a necessity in any book that isn't just reams of numbers—I had to choose one explanation from a variety of possibilities. Is there some force besides age behind what I call Wooderson's law (the fact that straight men of all ages are most interested in twenty-year-old women)? Perhaps. But I think it is very unlikely. "Correlation does not imply causation" is a good thing for everyone to keep in mind—and an excellent check on narrative overreach. But a snappy phrase doesn't mean that the question of causation isn't itself interesting, and I've tried to attribute causes only where they are most justified.

For almost all the parts of *Dataclysm* that overlap with posts on OkCupid's blog, I chose to redo the work from scratch, on the most recent data, rather than quote my own previous findings. I did so because, frankly, I wanted to double-check what I'd done. The research published there from 2009 through 2011 was put together piecemeal. Many different people—I can count at least five—had pulled male-female message-reply rates for me over those three years, just to name one frequently used data point, and going back through my records of this data, there was no way to be sure what data set had generated the results. Doing it again myself, I could be sure. I could also enforce a uniform standard across all my research (for example, restricting analysis to only people ages twenty to fifty—a choice I made because those are the ages where I knew I had representative data).

Because the research is new, the numbers printed in *Dataclysm* are different from the numbers on the blog. Curves bend in slightly new ways. Graphs are a bit thicker or perhaps a bit thinner in places. The findings in the book and on the blog are nonetheless consistent. Ironically, with research like this, precision is often less appropriate than a generalization. That's why I often round findings to the nearest 5 or 10 and the words "roughly" and "approximately" and "about" appear frequently in these pages. When you see in some article that "89.6 percent" of people do *x*, the real finding is that "many" or "nearly all" or "roughly 90 percent" of them do it, it's just that the writer probably thought the decimals sounded cooler and more authoritative. The next time a scientist runs the numbers, perhaps the outcome will be 85.2 percent. The next time, maybe it's 93.4. Look out at the churning ocean and ask yourself exactly which whitecap is "sea level." It's a pointless exercise at best. At worst, it's a misleading one.

If you trace the findings in *Dataclysm* back to the original sources, the OkCupid data isn't the only place you'll see discrepancies. This data of our lives, being itself practically a living thing, is always changing. For example, my Klout score, which is holding steady at 34 as I write these words, will have no doubt gone up by the time you read them, since part of my obligation to Crown will be to tweet about this book. User engagement, ho!

Sometimes the numbers shift for no obvious reason. My copy editor and I had a mess of a time pinning down the Google autocompletes for prompts like "Why do women . . ." Google had given each of us slightly different results (". . . wear thongs?" was my third result to the above, presumably because that's a typically male question [?]. Hers was ". . . wear bras?"). Then when I checked a few weeks later, I myself saw something different: ". . . wear high heels?" Since it was the most recent result, that's what ended up in the book.

As interesting a tool as it is, the black box of Google's autocomplete (and Google Trends, for that matter) is an example of one of the worst things about today's data science—its opaqueness. Corroboration, so important to the scientific method, is difficult, because so much information is proprietary (and here OkCupid is as guilty as anyone). Even as most social media companies trumpet the hugeness and potential of their data, the bulk of it has stayed off-limits to the larger world. Data sets currently move through the research community like yeti—*I have a bunch of interesting stuff but I can't say from where; I heard someone at Temple has tons of Amazon reviews; I think L has a scrape of Facebook*. That last is something I was told by three unrelated academics; they referred to another scientist by name, which I've here obscured. L does in fact have that rogue Facebook scrape—I met him and confirmed—but he can't show it to anyone. He's really not supposed to have it at all. Data is money, which means companies treat it as such—and though some digital data sits out in the open, it's secured behind legal walls as thick as any vault's. If you look at your friend Lisa's Facebook page, observe that her name is Lisa, and publish that fact (anywhere!)—you have technically stolen Facebook's data. If you've ever signed up for a website and given a fake zip code or a fake birthday, you have violated the Computer Fraud and Abuse Act. Any child under thirteen who visits newyorktimes.com violates their Terms of Service and is a criminal—not just in theory, but according to the working doctrine of the Department of Justice.* The examples I've laid out are extreme, sure, but the laws involved are so broadly written as to ensure that, essentially, every Internet-using American is a tort-feasing felon on a lifelong spree of depraved web browsing. Whether anyone penalizes you for your "crime" is another matter, but, legally, you are prostrate, a boot on your neck. A company's general counsel, or a district attorney looking to please an important corporate donor, can destroy your life simply by deciding to press. When it suits, they do. So social scientists are very

* For more on the Kafkaesque implications of the CFAA, please see "Until Today, If You Were 17, It Could Have Been Illegal to Read Seventeeen.com Under the CFAA" and "Are You a Teenager Who Reads News Online? According to the Justice Department, You May Be a Criminal," both published by the Electronic Frontier Foundation.

cagey with data sets; actually, more than yeti, they treat them like big bags of weed—possessive, slightly paranoid, always curious who else is holding and how dank that shit is.

Increasingly the preferred practice is to bring researchers in-house rather than release information outside.* And that approach has yielded, among many fruits, the novel research by Facebook's data team and Seth Stephens-Davidowitz's fine work at Google, both of which I've drawn on here. I hope more companies follow this model, and that eventually we, the owners of the sites, will find a way to release our data for the public good without jeopardizing our users' privacy in the act.

∞

It's old hat now, but the app Shazam was, to me, one of the first great wonders of the iPhone. It's a little program for identifying music—if some song is playing, and you want to know what it is, you just turn on the app and hold up your phone. Shazam listens through the microphone, and, like, two seconds later, it tells you what you're listening to. The first time someone did it in front of me, I was just blown away, not only at how little the software needed to get the song right (it can often work through walls or above the din of a bar), but at how fast it worked. It was the closest thing I'd seen to magic, at least until I came to know a certain able necromancer who, at a whim, could summon fees and add them to my goddamn kitchen renovation. But anyway, as I later found out, Shazam relies on an incredible principle: that almost any piece of music can be identified by the up/down pattern in the melody—you can ignore everything else: key, rhythm, lyrics, arrangement . . . To know the song, you just need a map of the notes' rise and fall. This melodic contour is called the song's Parsons code, named after the musicologist who developed it in the 1970s. The code for the first two lines of "Happy Birthday" is •RUDUDDRUDUD, with U meaning "melody up," D meaning "melody down," and R for "re-

* I wish this were called hotboxing, but sadly, no.

peated note." The dot • just marks the beginning of the tune, which of course isn't up or down from anything. Hum it to yourself to check:

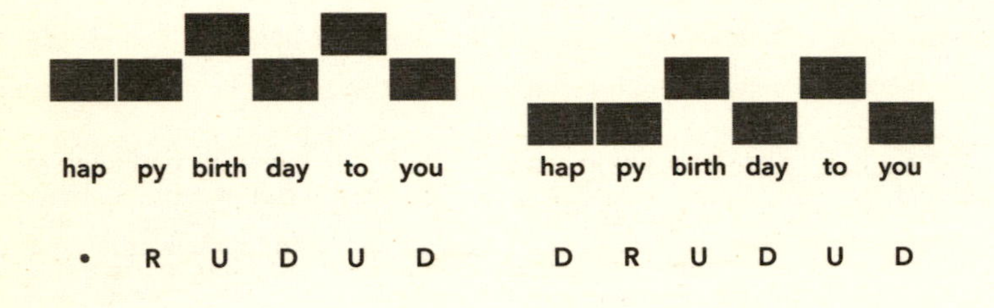

As crazy as it seems, the code for "Happy Birthday" is practically unique across the entire catalog of recorded music, as is the code for almost all songs. And it's because these few letters are such a concise description that Shazam is so fast: instead of a guitar, Paul McCartney, and just the right amount of reverb, "Yesterday" starts with •DRUUUUUUDDR. That's a lot easier to understand.

Like an app straining for a song, data science is about finding patterns. Time after time, I—and the many other people doing work like me—have had to devise methods, structures, even shortcuts to find the signal amidst the noise. We're all looking for our own Parsons code. Something so simple and yet so powerful is a once-in-a-lifetime discovery, but luckily there are a lot of lifetimes out there. And for any problem that data science might face, this book has been my way to say: I like our odds.

Notes

We no longer live in a world where a reader depends on endnotes for "more information" or to seek proof of facts or claims. For example, I imagine any reader interested in Sullivan Ballou will have Googled him long before she consults these notes and transcribes into her browser the links I've provided. So I have used this section to focus on the many sources that have contributed not only facts but ideas to this book. I've also used it to substantiate or explain claims about my own proprietary data.

Since the subject of *Dataclysm* is changing almost daily, I've decided to enhance this section online at dataclysm.org/endnotes, where you will find additional source material and findings from emerging research.

Introduction

15 ***10 million people will use the site*** For this number, I counted every person who logged into OkCupid in the twelve months trailing April 2014: 10,922,722.

16 ***Tonight, some thirty thousand couples*** It's the great unknowable of running an online dating site: How many of the users actually meet in person? And what happens next? This passage represents my best guesses at some basic in-person metrics. I used two separate methods:

1. I assumed someone who's actively using OkCupid goes on one date every other month. I think this is conservative. At roughly 4,000,000 active users each month, that means roughly 65,000 people go on dates each day, meaning roughly 30,000 couples.

2. Every day 300 couples wind their way through our "account disable" interface to let us know that they no longer need OkCupid specifically because they have found a steady relationship on OkCu-

pid. These are couples who (a) are dating seriously enough to shut down their OkCupid accounts, and who (b) are willing to go through the trouble of filling out a bunch of forms to let us know their new relationship status. I estimate that Group B represents only 1 in 10 of the long-term couples actually created by the site. And I estimate that Group A represents the outcome of only 1 in 10 first dates. Therefore, there must be 3,000 long-term couples, from 30,000 first dates each day. Of every 3,000 long-term couples, I believe something less than 1 in 10 go on to get married. One way to look at this: How many serious relationships did you have before you found the person you settled down with? I imagine the average number is roughly 10.

These appraisals together are mutually supporting, at least of the "first dates" number, and even if it's approximate, I think the deeper metrics follow plausibly.

21 ***ratings of pizza joints on Foursquare*** Ratings from a random sample of 305 New York City pizza places accessed through Foursquare's public API.

22 ***the recent approval ratings for Congress*** These were collected from the 529 polls measuring "congressional job approvals" listed on the site real clearpolitics.com from January 26, 2009, through September 14, 2013. See realclearpolitics.com/epolls/other/congressional_job_approval-903.html#polls.

22 ***NBA players by how often*** The chart shows percent of games started for each of the players listed on a team roster for the 2012–2013 season on espn.com. Yes, I'm counting the 76ers as an NBA team.

23 ***6 percent*** This number comes from taking the geometric mean of the distances between each of the 21 discrete data points along the curves. So, for curves *a* and *b*, I calculated:

$$\sqrt{\sum_{k=1}^{21}(a_k - b_k)^2}$$

Which equals 0.056.

24 ***58 percent of men*** The male attractiveness curve is centered more than a whole standard deviation below the female. Translating the same disparity to IQ means that the median male IQ would be slightly lower than 85, which is the threshold for "borderline intellectual functioning." For example, the US Army doesn't accept applicants with IQs below 85. I say "brain damaged" as a bit of hyperbole meant to capture this shift. Strictly speaking, I mean that 58 percent of men would have IQs lower than 85.

24 ***half the single people in the United States*** Specifying the reach of the dating data I have was a challenge. I've strived to do so in broad, easy-to-grasp terms because, unlike Facebook or Twitter, I know much of my reading audience has never used a dating site. If you've been married or in a relationship since the late '90s or before, you have never needed online dating. According to the 2011 Census numbers, there are 103 million single people ages fifteen to sixty-four in the United States—that counts *everyone* who isn't legally married, including many people who are actually in long-term relationships and nearly every gay person. Together, Tinder, OkCupid, DateHookup, and Match.com registered 57 million US accounts from 2011 to 2013, and 23 million in the last of those three years alone. "Half" is my approximation of 57/103, minus the 10 to 15 percent wastage in overlap and duplicate accounts.

24 ***"Women are inclined to regret"*** This quote is from the "Findings" section of the February 2014 issue of *Harper's* by Rafil Kroll-Zaidi.

24 ***A beta curve plots*** My data researcher, Tom Quisel, helped me put the binomial nature of beta curves into simple terms. He also pointed out that they're used to model weather, and ran the comparisons to the by-city patterns on weatherbug.com.

27 ***Some 87 percent of the United States is online*** See Susannah Fox and Lee Rainie, "Summary of Findings," Pew Research Internet Project, Pew Research Center, February 27, 2014, pewinternet.org/2014/02/27/summary-of-findings-3.

27 ***that number holds . . .*** For example, Internet use among white, African American, and Hispanic Americans is 85, 81, and 83 percent, respectively. One can only assume adoption among Asian Americans is similar. Adoption is above 80 percent for all age groups, save people sixty-five and older. Susannah Fox and Lee Rainie, "Internet Users in 2014," Pew Research Internet Project, Pew Research Center, February 27, 2014, pewinternet.org/files/2014/02/12-internet-users-in-2014.jpg.

27 ***More than 1 out of every 3 Americans access Facebook*** Facebook reported 128 million US users in August 2013. Facebook had at least 1.26 billion users worldwide in September 2013. World and US population statistics are from Wikipedia. See expandedramblings.com/index.php/by-the-numbers-17-amazing-facebook-stats.

27 ***fundamentally populist*** This is something like common knowledge among people who study social media adoption beyond the Google Glasshole/Technocrat use case. See Pew Research Center's "Demographics of Key Social Networking Platforms" (2013). The report shows no statistically significant difference in rates of Twitter use between the "high school grad or less" and "College +" educational cohorts (coming in at 17 percent and 18 percent, respectively). Pew surveys a random cross-section of Americans eighteen years old or older, so very few of the "high school grad or less" cohort are that way simply because they're still *in* high school. By ethnicity, Pew reports adoption rates of 29 percent among blacks and 16 percent among both whites and Hispanics. The full report, by Maeve Duggan and Aaron Smith, is here: pewinternet.org/2013/12/30/demographics-of-key-social-networking-platforms.

28 ***It's called WEIRD research*** This fact and my general take on the phenomenon are adapted from "Psychology Is WEIRD," by Bethany Brookshire, in *Slate*. See also "The Roar of the Crowd," *The Economist*, May 24, 2012, economist.com/node/21555876.

30 ***Pharaoh Narmer*** As you can imagine, this is up for debate, though Narmer, also known as Serket, is a defensible choice. In earlier drafts I had Gilgamesh, the Akkadian hero, in this place because J. M. Roberts, in his *History of the World* (New York: Oxford University Press, 1993), chooses Gilgamesh. I eventually went with Narmer because his life is dated several centuries earlier, and he seemed to me as likely to have actually lived. Yahoo! Answers also mentions Elvis Presley.

Chapter 1: Wooderson's Law

41 ***This isn't survey data*** This is a good place to point out that for anyone's attractiveness to have been considered in my analysis in this book, that person needed to have received votes from at least twenty-five other people. For something as idiosyncratic as attraction, I felt an average score comprising fewer than twenty-five votes wasn't reliable.

47 ***per the US Census*** These numbers are from the US Census Bureau's "Marital Status of People 15 Years and Over, by Age, Sex, Personal Earnings, Race, and Hispanic Origin, 2011."

Chapter 2: Death by a Thousand Mehs

54 ***"Beauty is looks you can never forget"*** John Waters, *Shock Value: A Tasteful Book About Bad Taste* (Philadelphia: Running Press, 2005), p. 128.

56 ***concept called* variance** I used standard deviation to measure variance throughout this chapter.

59 ***the "pratfall effect"*** A Google search for "pratfall effect" will yield many examples. I particularly relied on the précis "The Positive Effect of Negative Information" by Bill Snyder and the original paper he summarizes, "When Blemishing Leads to Blossoming: The Positive Effect of Negative Information," by Danit Ein-Gar, Zakary Tormala,

and Shiv Tormala, *Journal of Consumer Research* 38, no. 5 (2012): 846–59.

59 ***Our sense of smell*** For this passage, I relied on Fabian Grabenhorst et al., "How Pleasant and Unpleasant Stimuli Combine in Different Brain Regions: Odor Mixtures," *Journal of Neuroscience* 27, no. 49 (2007): 13532–40, doi: 10.1523/JNEUROSCI.3337–07.2007. Wikipedia's "Indole" entry describes its "intense fecal smell." For more on indole's role in perfumes and in naturally occurring flower scents, see, as I did, perfumeshrine.blogspot.com/201%5/jasmine-indolic-vs-non-indolic.html.

59 ***On the following page are six women*** We received these permissions using a double-blind system, to protect user privacy. I submitted criteria (women, high variance scores, midrange overall attractiveness) to OkCupid's data team. The data team generated a list of possible names, which they passed on to our admin. She then had a list of names, with no other information attached, and was told to contact them for blanket photo authorization. (We commonly receive press requests for user photos, so this type of outreach isn't unusual.) A photo and its unique attributes were only connected once permission was granted.

Chapter 3: Writing on the Wall

65 ***Nostalgia used to be called*** Because the phenomenon is so interesting (and unexpected) and one link leads to another, my sources for this passage were many. These I drew on directly:

"Dying to Go Home," by Jackie Rosenhek, *Doctor's Review*, December 2008, doctorsreview.com/history/dying-to-go-home.

"Beware Social Nostalgia," by Stephanie Coontz, *New York Times*, May 19, 2013, nytimes.com/2013/05/19/opinion/sunday/coontz-beware-social-nostalgia.html.

"When Nostalgia Was a Disease," by Julie Beck, *The Atlantic*, August

2013, theatlantic.com/health/archive/2013/08/when-nostalgia-was-a-disease/278648.

The "Nostalgia" entry on qi.com: qi.com/infocloud/nostalgia.

65 ***people under eighteen aren't using Facebook*** The earnings call in question reviewed Facebook's fourth-quarter performance, 2013. See Joanna Stern, "Teens Are Leaving Facebook and This Is Where They Are Going," ABCNews, October 31, 2013, abcnews.go.com/story?id=20739310.

66 ***Major Sullivan Ballou*** The basic facts surrounding the letter can be found here: pbs.org/civilwar/war/ballou_letter.html. Though the letter was never mailed, it was included with Ballou's belongings and returned to his family after his death.

67 ***There will be more words written on Twitter*** I calculate this as follows: 129,864,880 books have been written, at least according to Google. That number is laughably precise; however, given that they have already logged 30 million of them, and indexing things is their business, their guess should be considered a plausible estimate. See Ben Parr, "Google: There Are 129,864,880 Books in the Entire World," *Mashable*, August 5, 2010, mashable.com/201%8/05/number-of-books-in-the-world.

According to Amazon, the median length of a novel is 64,000 words. Since it's very likely that the median and mean are close here, I'm comfortable using it as an average. I don't think novels are necessarily longer or shorter than other books. See Gabe Habash, "The Average Book Has 64,500 Words," *PWxyz*, March 6, 2012, blogs.publishersweekly.com/blogs/PWxyz/2012/03/06/the-average-book-has-64500-words.

These two numbers together yield 8,311,352,320,000 words ever in print.

Twitter reported 500 million tweets a day in August 2013. See blog.twitter.com/2013/new-tweets-per-second-record-and-how.

I estimate that each tweet has 20 words. So at 10 billion words a day, it will take Twitter 831 days (2.3 years) to surpass all of printed literature in volume. This is obviously meant to be an approximation, and a conservative one at that. In all likelihood, Twitter will do it much faster, since the rate of tweets per day is increasing rapidly.

67 ***"You only have to look on Twitter"*** Mr. Fiennes's quote was covered extensively. See Lucy Jones, "Ralph Fiennes Blames Twitter for 'Eroding' Language," *Telegraph*, October 27, 2012, telegraph.co.uk/technology/twitter/8853427/Ralph-Fiennes-blames-Twitter-for-eroding-language.html.

67 ***Even basic analysis shows*** Here and in all my own Twitter analysis I use the tweets and followers generated by a representative corpus of 1.2 million accounts, collected at random by my research team.

67 ***The OEC is the canonical census*** More on the OEC and its most common words can be found here: en.wikipedia.org/wiki/Most_common_words_in_English.

The OEC lists only *lemmas*—that is, the base word root of a related lexical pattern. For example, it counts *have* for *had, having, has*, and so on. I chose not to do this in my Twitter research. Though my choice makes comparing the lists directly more difficult, I preferred to present the data in as raw a state as possible.

69 ***Mark Liberman*** Professor Liberman's blog Language Log (languagelog.ldc.upenn.edu/nll) contains a trove of interesting textual analysis. See "Up in UR Internets, Shortening All the Words," October 28, 2011, languagelog.ldc.upenn.edu/nll/?p=3532, for his discussion of the Fiennes quote in particular.

70 ***A team at Arizona State*** The Twitter textual analysis in the rest of this paragraph is drawn from "*Dude, srsly?*: The Surprisingly Formal Nature of Twitter's Language," by Yuheng Hu, Kartik Talamadupula, and Subbarao Kambhampati, paper presented at the seventh annual International AAAI Conference on Weblogs and Social Media,

Cambridge, Massachusetts, July 8–11, 2013, aaai.org/ocs/index.php/ICWSM/ICWSM13/paper/view/6139.

71 ***Here I've excerpted an early attempt*** The table and the subsequent discussion of the word "tribes" on Twitter are drawn from "Word Usage Mirrors Community Structure in the Online Social Network Twitter," by John Bryden, Sebastian Funk, and Vincent AA Jansen, *EPJ Data Science* 2, no. 3 (2013). I also draw from their "Additional Material" containing raw community word lists not used in the paper itself. The full paper, along with links to the additional material, can be found here: epjdatascience.com/content/2/1/3.

72 ***This body of data has created a new field*** This method of mining Google Books for cultural trends was first proposed in *Science* in the article "Quantitative Analysis of Culture Using Millions of Digitized Books," by Jean-Baptiste Michel et al., *Science* 331, no. 6014 (2011): 176–82, doi:10.1126/science.1199644.

My graph of food words over time is a reproduction of their exploration of the same terms in that paper. My graph of year words over time is an adaptation of their method, rather than a reproduction. The paper references a "half-life" of memory that I was not able to reproduce. Nonetheless, the writers' claim that "We are forgetting our past faster with each passing year" is clearly directionally correct. The paper has much more of interest than just the two charts I've referenced here and is worth reading in full.

76 ***Below is a scatter chart of 100,000 messages*** No private messages were read by anyone in performing this analysis. The number of keystrokes and typing time are logged automatically for a sample of OkCupid's users as part of our ongoing spam-detection software. Since I didn't read any actual user messages, the quoted text of the three-letter message "hey" is a likelihood rather than a certainty. About 80 percent of three-letter messages on the site are "hey." "Sup" is the next most popular, then "wow." Given the overwhelming popularity

of "hey," and that I was making a joke, and that any of the alternatives would've worked just as well, I was comfortable picking "hey" in this context.

78 ***"I'm a smoker too"*** This private message, presented verbatim and complete, came to my attention in a context outside this book, and I received the sender's permission to both reprint and discuss it here.

Chapter 4: You Gotta Be the Glue

84 ***"social graphs"*** The network plots on pages 74 and 75 were generated by James Dowdell, using the same general graphic scheme used by Lars Backstrom and Jon Kleinberg in their paper "Romantic Partnerships and the Dispersion of Social Ties: A Network Analysis of Relationship Status on Facebook," presented at the 18th ACM Conference on Computer-Supported Cooperative Work and Social Computing, Baltimore, Maryland, February 15–19, 2014, delivery .acm.org/10.1145/2540000/2531642/p831-backstrom.pdf.

85 ***I spent years touring in a band*** My band is called Bishop Allen; Justin Rice is the band's other half. You can find our songs on Spotify, or on the nearest torrent, or on iTunes. For anyone interested, my personal recommendations are the songs "Like Castanets," "Click Click Click Click," "Chinatown Bus," "Start Again," and "Little Black Ache."

86 ***In 1735, Leonhard Euler*** Though I was familiar with Euler, the bridges problem, and their role in the genesis of graph theory from my time as a math major, I relied on Wikipedia's "Seven Bridges of Königsberg" entry for the minutiae surrounding the problem and its solution.

86 ***has since helped us understand*** A good resource for both classic and modern uses of graph theory is here: world.mathigon.org/Graph_Theory.

87 ***Stanley Milgram*** Like Euler, Milgram and his work have been familiar to me for years. However, I relied on his Wikipedia entry for the details of his "Six Degrees" experiment.

87 ***Facebook allowed us to see*** See "The Anatomy of the Facebook Social Graph," by Johan Ugander et al. (arXiv preprint, 2011, arXiv: 1111.4503).

87 ***Pixar famously put*** The idea was Steve Jobs's. I first heard of this anecdote in Jonah Lehrer's *Imagine* (Edinburgh, UK: Canongate, 2012). See BuzzFeed's "Inside Steve Jobs' Mind-Blowing Pixar Campus," by Adam B. Vary, for more details. Vary mind-blowingly interviews Craig Payne, a senior Pixar manager: buzzfeed.com/adambvary/inside-steve-jobs-mindblowing-pixar-campus.

87 ***"the strength of weak ties"*** See "The Strength of Weak Ties" by Mark S. Granovetter, *American Journal of Sociology* 78, no. 6 (1973): 1360–80.

87 ***Another long-held idea in network theory*** Though embeddedness was first proposed by Granovetter in 1985, my remaining discussion of embeddedness and of interpersonal network theory is drawn from the primary source behind this chapter, Backstrom and Kleinberg's "Romantic Partnerships." I apply their heuristic to my own networks and somewhat simplify their original work for a nonacademic audience.

89 ***an astounding 75 percent of the time*** Backstrom and Kleinberg define many subtly different mathematical kinds of dispersion. My number here refers to the accuracy they reported with the method they call "recursive dispersion."

90 ***50 percent more likely*** This is drawn from the following passage in Backstrom and Kleinberg's paper: "We find that relationships on which recursive dispersion fails to correctly identify the partner are significantly more likely to transition to 'single' status [that is, break up] over a 60-day period. This effect holds across all relationship ages and is particularly pronounced for relationships up to 12 months in age; here the transition probability is roughly 50% greater when recursive dispersion fails to recognize the partner."

Chapter 5: There's No Success Like Failure

94 ***one of Google's best designers*** Douglas Bowman leaving Google is a famous event in tech circles. See his own post "Goodbye, Google" at stopdesign.com/archive/2009/03/20/goodbye-google.html.

96 ***no evidence of people gaming the system*** It was fairly simple to unscramble a Crazy Blind Date photo; we knew this would be the case. Sure enough, about a week after launch a few hackers had built apps to de-anonymize the photos. However, these apps never caught on, mostly because they were difficult to use and even then only worked part of the time. These unscramblers were not a factor in Crazy Blind Date's product trajectory or the data it generated. The scrambled example photo printed in the book is a stock photo, licensed from Getty Images.

Chapter 6: The Confounding Factor

107 ***of a certain type*** See, for example, "Blacks Still Dying More from Cancer Than Whites," by Jordan Lite, *Scientific American*, February 2009. Also see the Sentencing Project's "Criminal Justice Primer for the 111th Congress," which details many depressing disparities in the sentences handed down to whites, compared to minority defendants: sentencingproject.org/doc/publications/cjprimer2009.pdf.

108 ***conclusions like this*** The headline cited is from ThinkProgress.org. "Study: Black Defendants Are at Least 30% More Likely to Be Imprisoned Than White Defendants for the Same Crime," by Inimai Chettiar, August 30, 2012, thinkprogress.org/justice/2012/08/30/770501/study-black-defendants-are-at-least-30-more-likely-to-be-imprisoned-than-white-defendants-for-the-same-crime.

108 ***in the 97,000 results*** It's a bit of a hack to get Google to give you a number here. My exact query was for " 'black quarterback' –adsffsdada." Using the minus sign with the nonsense word keeps the page from automatically returning images instead of the "about

97,000 results" text. I'm sure without the browser in front of you, this all sounds mystifying. Try it yourself if you care, and you'll see immediately what I mean. Also, this is another example of a raw number that has changed during the course of writing this book. I've also gotten "89,800 results" returned to me.

108 ***I found only one article*** See Jason Lisk, "Quarterbacks and Whether Race Matters," *The Big Lead*, December 2, 2010, thebiglead.com/2010/12/02/quarterbacks-and-whether-race-matters. Of course, the fact that I found only one writer who calculates quarterback rating by race is hardly proof that no other writer has made the calculation. However, I spent several hours combing results and found only Lisk.

109 ***the four largest racial groups*** 15 percent of OkCupid users who select an ethnicity select more than one race; 3 percent select a race other than the four largest. These people are excluded from the analysis, as are people who neglected to choose a race at all.

110 ***"normalize" each row*** I normalized against the simple average in each row, rather than the weighted average. Because of the preponderance of white people, the latter technique would've skewed the matrix, functionally using what everyone thinks of white people as the "norm." A simple average captures the following: "When a person of race A meets an arbitrary person of race B, how does A appraise B, relative to A's appraisals of other races?" That's the interesting question, and what we want to investigate.

112 ***There is no cadre of racists*** An analysis of individual bias applied by non-black men to black female profiles shows a median deduction of 0.6 stars, with most of the sample applying a deduction from 0.2 to 1.0 stars. 82 percent of the sample shows at least some consistent anti-black bias.

112 ***Here are our numbers*** Though the numbers I list for OkCupid here were generated from internal data, you can see those numbers corroborated and compared to Quantcast's national averages by visiting https://

www.quantcast.com/okcupid.com?country=US. Select "Ethnicity" from the Demographics menu and expand the "US average" feature.

118 ***OkCupid users putting it in their own words*** These excerpts are from user-submitted "Success Stories" published on the site. Bella and Patrick's is here: https://www.okcupid.com/success/story?id=2855. Dan and Jenn's is here: https://www.okcupid.com/success/story?id=2587.

119 ***"There are very few"*** Barack Obama's quote is excerpted from his comments on the George Zimmerman verdict: whitehouse.gov/the-press-office/2013/07/19/remarks-president-trayvon-martin.

119 ***One paper asked*** See "Are Emily and Greg More Employable Than Lakisha and Jamal? A Field Experiment on Labor Market Discrimination," by Marianne Bertrand and Sendhil Mullainathan, *American Economic Review* 94, no. 4 (2004): 991–1013, doi: 10.1257/0002828042002561.

121 ***Osagie K. Obasogie*** My discussion of Obasogie's work relies on Francie Latour's *Boston Globe* article "How Blind People See Race," January 19, 2014. Latour provides a précis of Obasogie's book *Blinded by Sight: Seeing Race Through the Eyes of the Blind* (Redwood City, CA: Stanford University Press, 2014), and interviews him.

122 ***Baywatch*** I was in Japan in 1992. *Baywatch* was popular worldwide by then, but didn't arrive in the Japanese mainstream until a year later. Nonetheless, surf culture, California, and sun-kissed blondness were already everywhere. When you walked into a "cool" clothing store, they'd be playing the Beach Boys. In 1992. Stuff like "Surfin' Safari," not "Kokomo."

Chapter 7: The Beauty Myth in Apotheosis

127 ***Korean proverb*** I got this from William Manchester's biography of Douglas MacArthur, *American Caesar* (New York: Little, Brown, 1978), which, in the death throes of this book, I was reading to get my mind off data.

128 ***beauty operates on a Richter scale*** I was already familiar with the logarithmic nature of the Richter scale, but relied on the Wikipedia entry for "Richter magnitude scale" to understand the implications of the benchmark magnitudes. In comparing beauty to the scale, I am, of course, employing a bit of poetic license; the functions are not exactly the same.

129 ***Here is data for interview requests*** The Shiftgig data was provided by their data team and with the gracious cooperation of founder Eddie Lou.

129 ***And for friend counts*** These are the aggregated and anonymized friend counts for OkCupid users who've elected to connect their OkCupid accounts to their Facebook accounts.

130 ***a foundational paper of social psychology*** See "What Is Beautiful Is Good," by Karen Dion, Ellen Berscheid, and Elaine Walster in *Journal of Personality and Social Psychology* 24 (1972): 285–90.

130 ***It was the first in a now long line . . .*** This passage adapts conclusions from and directly quotes "Pretty Smart? Why We Equate Beauty with Truth," by Robert M. Sapolsky, in the *Wall Street Journal*, January 17, 2014. The Duke neuropsychologists alluded to are Takashi Tsukiura and Roberto Cabeza. See also "Jurors Biased in Sentencing Decisions by the Attractiveness of the Defendant" at *Psychology and Crime News* for an overview of the effects of physical attractiveness in the criminal justice process: crimepsychblog.com/?p=1437, posted by user EmmaB, April 3, 2007.

134 ***both Tumblr and Pinterest*** See "A New Policy Against Self-Harm Blogs," Tumblr's staff blog, March 1, 2012, staff.tumblr.com/post/18132624829/self-harm-blogs.

See also "Pinterest 'Thinspiration' Content Banned According to New Acceptable Use Policy," by Ellie Krupnick, *Huffington Post*, March 26, 2012, huffingtonpost.com/2012/03/26/pinterest-thinspiration-content-banned_n_1380484.html.

The *Huffington Post* has actively covered the "thinspiration" phenomenon. See "The Hunger Blogs: A Secret World of Teenage 'Thinspiration,' " by Carolyn Gregoire, February 8, 2012, huffingtonpost.com/2012/02/08/thinspiration-blogs_n_1264459.html.

For more on "thighgap" (and for evidence that altering the Terms of Service did not solve the problem), see "The Sexualization of the Thigh Gap," by Allie Jones, on *The Wire*, November 22, 2013, thewire.com/culture/2013/11/sexualization-thigh-gap/355434.

Chapter 8: It's What's Inside That Counts

137 ***That's been the popular standard since*** These basic facts on the origins of Gallup were found on the "Gallup (company)" Wikipedia entry.

137 ***surveys have historically*** As I mention in the text and in the footnotes to this chapter, the idea of using Google Trends to look at taboos is the brainchild of Seth Stephens-Davidowitz. His June 9, 2012, article in the *New York Times*, "How Racist Are We? Ask Google," and his 2013 Harvard PhD dissertation, "Essays Using Google Data," http://nrs.harvard.edu/urn-3:HUL.InstRepos:10984881, were the inspiration for this chapter. For the question of exactly how much Obama's race cost him in the 2008 election, picked up later in the chapter, I rely directly on Stephens-Davidowitz's work. For the over-time use of the word "nigger" and in the other direct citations of Google Trends findings in the chapter, the work is my own, though I am adapting a method he first suggested.

Though Stephens-Davidowitz now works at Google, I emphasize that his search research is always based on public and anonymous sources, not on privileged access to anyone's personal search history. My own search research is similarly based on a public, anonymous source, namely Google Trends: google.com/trends.

137 ***This tendency is called*** I used Wikipedia's "Social desirability bias" entry as my source for basic details here.

137 ***The most famous case*** The Bradley effect first came to my attention during the 2008 campaign, as pundits wondered how it would affect Obama's polling on Election Day. Here, I relied on the Wikipedia entry "Bradley effect" for basic facts surrounding Tom Bradley's defeat.

138 ***Since the service launched*** See Nick Bilton, "Google Search Terms Can Predict Stock Market, Study Finds," *New York Times* Bits blog, April 26, 2013. See also Casey Johnston, "Google Trends Reveals Clues About the Mentality of Richer Nations," *Arstechnica*, April 5, 2012, arstechnica.com/gadgets/2012/04/google-trends-reveals-clues-about-the-mentality-of-richer-nations/; and Tobias Preis et al., "Quantifying the Advantage of Looking Forward," *Scientific Reports* 2, no. 350 (2012), doi: 10.1038/srep00350.

138 ***track epidemics of flu*** Google Flu was first developed in the paper "Detecting Influenza Epidemics Using Search Engine Query Data," by Jeremy Ginsberg et al. in *Nature* 457 (2009): 1012–14, doi:10.1038/nature07634. Recently, Flu's efficacy has been found wanting: see Kaiser Fung, "Google Flu Trends' Failure Shows Good Data > Big Data," *Harvard Business Review* Blog Network, March 25, 2014.

138 ***included in 7 million searches a year*** Stephens-Davidowitz, "How Racist Are We?"

139 ***more American than "apple pie"*** Google Trends index for US searches, January 2004–September 2013, for "apple pie": 25. For "nigger": 32.

139 ***And, tellingly*** The ratio of "nigga":"nigger" is thirty times higher in tweets sent from my Twitter corpus than reflected in Google Trends. That is, on Twitter "nigger" appears thirty times less frequently.

140 ***roughly 1 in 100 searches for "Obama"*** Stephens-Davidowitz shared this fact with me over e-mail.

140 ***25 percent below the pre-Obama status quo*** Stephens-Davidowitz, "How Racist Are We?" This is also confirmable firsthand through Google Trends.

141 ***Other awful terms*** These racial epithets are far less common on Twitter, in private messages to OkCupid, and in Google search, as confirmed by Stephens-Davidowitz via e-mail.

141 ***If you're not familiar with autocomplete*** The algorithm that supplies Google autocomplete is the blackest of the black boxes. There is little definitive information on how it works. Danny Sullivan at searchengineland.com offers a thorough, if mostly ad hoc, overview at searchengineland.com/how-google-instant-autocomplete-suggestions-work-62592. Because autocomplete seems to factor in your personal search history, individual results are highly variable here. If you try to replicate my results for yourself, make sure to use an "Incognito" session of Chrome, as I did, so that Google has no prior personal data to work with. If you're a Safari user, select "Private Browsing."

142 ***one such result*** See Paul Baker and Amanda Potts, " 'Why Do White People Have Thin Lips?' Google and the Perpetuation of Stereotypes Via Auto-Complete Search Forms," *Critical Discourse Studies* 10, no. 2 (2013): 187–204.

142 ***Go to your search bar with*** This long string of queries was suggested to me by Sean Mathey, on the van ride home following a camping trip where we played a lot of Magic: the Gathering.

143 ***I'll let Republican strategist Lee Atwater explain*** See Rick Perlstein, "Exclusive: Lee Atwater's Infamous 1981 Interview on the Southern Strategy," *The Nation*, November 13, 2012, thenation.com/article/170841/exclusive-lee-atwaters-infamous-1981-interview-southern-strategy. Original quote from Alexander P. Lamis's book *The Two-Party South* (New York: Oxford University Press, 1984), via Wikiquote's "Lee Atwater" entry.

144 ***Consider two media markets*** Stephens-Davidowitz, "How Racist Are We?"

145 ***In my opinion, Muhammad Ali*** I read David Remnick's *King of the World* (New York: Random House, 1998) in 1999 and have admired Ali since. I verified certain basic facts surrounding Ali's Vietnam protest using his Wikipedia entry. For Ali's famous quote on the Viet Cong, I went with the popular and much more pithy misquotation of his actual words, which were, "My conscience won't let me go shoot my brother, or some darker people, or some poor hungry people in the mud for big powerful America. And shoot them for what? They never called me nigger, they never lynched me, they didn't put no dogs on me, they didn't rob me of my nationality, rape and kill my mother and father ... Shoot them for what? How can I shoot them poor people? Just take me to jail." The misquotation is identical in spirit, yet so much shorter and so much better known, that I decided it was acceptable in place of the actual quote.

You can hear him say those words (the longer quote) himself in the YouTube video "Muhammad Ali on the Vietnam War-Draft" at https://www.youtube.com/watch?v=HeFMyrWlZ68. In that video, he seems to be speaking right after a fight, and his speech is slow and deliberate. Hear him speak much more fluently on the same topic two years later in "Muhammad Ali Interview with Ian Wooldridge (1969)" at https://www.youtube.com/watch?v=dLam_GiQ2Ww.

Chapter 9: Days of Rage

151 ***Safiyyah Nawaz tweeted a silly joke*** My sources for information on Safiyyah and for the tweets surrounding her ordeal were:

Neetzan Zimmerman, "Teen Posts Joke on Twitter, Internet Orders Her to Kill Herself," *Gawker*, January 2, 2013, gawker.com/1493156583.

Ryan Broderick, "Meet the 17-Year-Old Girl Who Stood Up to Death Threats After Her Tweet Went Viral on New Year's Eve," BuzzFeed, January 2, 2014, buzzfeed.com/ryanhatesthis/meet-the-17-year-old-girl-who-stood-up-to-death-threats-afte.

Ryan Broderick, "After Twitter Started Viciously Attacking Her over a Silly Joke, This Girl Handled It Like a Champ," BuzzFeed, January 2, 2014, buzzfeed.com/ryanhatesthis/after-twitter-started-attacking-her-over-a-silly-joke-this-g.

These articles put her retweet number at 14,000, but they were all published just a day later. My 16,000 was accurate as of mid-January 2014.

151 ***Katy Perry/Lady Gaga*** The counts of the retweets for their "Happy New Year" tweets were accurate as of mid-January 2014 and have most likely gone up somewhat in the time since.

152 ***comedian Natasha Leggero*** My sources for Leggero's joke and the subsequent uproar were:

" 'I'm Not Sorry': Comedian Natasha Leggero Refuses to Apologize Mocking Pearl Harbor Survivors on NBC," by that legendary gumshoe "DAILY MAIL REPORTER." *Mail Online*, January 4, 2014, dailymail.co.uk/news/article-2533809.

Ross Luippold, "Natasha Leggero's Stunning 'Not Sorry' Response over Controversial Pearl Harbor Joke," *Huffington Post*, January 4, 2014, huffingtonpost.com/2014/01/04/natasha-leggero-not-sorry-for-pearl-harbor-joke_n_4541354.html.

The derogatory tweets sent to Leggero were taken from a letter she published on her Tumblr: natashaleggero.com/letter.

153 ***Pictures of her family*** Justine's tweet and the outrage surrounding it were covered extensively. A decent overview of the uproar is here: "Justine Sacco: 5 Fast Facts You Need to Know," by Matthew Guariglia, on Heavy, December 21, 2013, heavy.com/news/2013/12/justine-sacco-iac-racist-pr-tweet-africa.

"This Is How a Woman's Offensive Tweet Became the World's Top Story," by Alison Vingiano, on BuzzFeed, is a more thorough survey, though one that conveniently omits BuzzFeed's own role in cheering on the mob: buzzfeed.com/alisonvingiano/this-is-how-a-womans-offensive-tweet-became-the-worlds-top-s.

"The Case of Justine Sacco and the Twitter Lynch Mob," by Sharon Waxman, in *The Wrap*, is a piece by someone who, like me, had worked with Justine: thewrap.com/case-justine-sacco-twitter-lynch-mob.

"Justine Sacco: How to Kill a Career with One Tweet," by Juana Poareo, is one of many pitiless articles, replete with screenshots of Justine's tweets in the aftermath. *The Guardian*, "Liberty Voice," December 22, 2013, guardianlv.com/2013/12/justine-sacco-how-to-kill-a-career-with-one-tweet.

A screenshot of Google's involvement in #HasJustineLandedYet can be found at "Justine Sacco Saga Sparks Criticism of Twitter Lynch Mob," by Lauren O'Neil, on CBCnews.com: cbc.ca/newsblogs/yourcommunity/2013/12/justine-sacco-saga-sparks-criticism-of-twitter-lynch-mob.html.

153 ***the Internet waited dry-mouthed*** Here, though there were many thousands of mean-spirited tweets to choose from in my data pull, I chose to print only tweets that had already been published by other sources:

@RonGeraci's tweet appears on his blog, The Minty Plum, in a thoughtful piece, "View from the Pitchfork Mob," January 12, 2014, the mintyplum.com/?p=486.

@noyokono's tweet appears in Frazier Tharpe, "PR Woman Tweets Racist Joke Before Flight, Twitter Waits for Her to Land and Get Fired," Complex.com, December 21, 2013,complex.com/pop-culture/2013/12/justine-sacco-racist-tweet.

@Kennymack1971's tweet appears in the Sharon Waxman article cited above, "The Case of Justine Sacco and the Twitter Lynch Mob."

154 ***her father isn't a billionaire*** Alec Hogg, "Rubbish Rumours. Tweeting Idiot Justine Sacco No Relation to Desmond Sacco, SA Min-

ing Billionaire," Biz News.com, December 27, 2013, biznews.com/tweeting-idiot-justine-sacco-no-relation-to-desmond-sacco-sa-mining-billionaire.

155 ***The reach of social media*** This research did not use our usual randomized Twitter corpus. We instead opted for a completist approach. For these numbers and the related chart, my team and I pulled every retweet of Safiyyah's joke and #HasJustineLandedYet. These numbers reflect our best estimates of who saw each.

156 ***Marine biologists*** Alan Yu, "More Than 300 Sharks in Australia Are Now on Twitter," *All Tech Considered*, December 31, 2013, NPR, npr.org/blogs/alltechconsidered/2013/12/31/258670211.

156 ***Rumors are mentioned*** My source for the history and science of rumors is Jesse Singal's piece "How to Fight a Rumor," *Boston Globe*, October 12, 2008, boston.com/bostonglobe/ideas/articles/2008/10/12/how_to_fight_a_rumor. The insight to connect rumors and social media virulence is his. He also quotes the "a man who lacks judgment . . ." passage from the Bible. "Judge not . . ." is my own addition, as is the "demon Rumor."

I also used "Rumor, Gossip and Urban Legends," by Nicholas DiFonzo and Prashant Bordia, in *Diogenes* 54, no. 1 (2007): 19–35, and Mr. DiFonzo's article "Rumour Research Can Douse Digital Wildfires" in *Nature* 493, no. 7431 (2013): 135.

158 ***a phenomenon first studied*** I was led to Suler's work from *Penny Arcade*. I drew basic facts on Suler and the online disinhibition effect from the Wikipedia entry for "Online disinhibition effect," which links to the comic. The comic itself is here: penny-arcade.com/comic/2004/03/19.

158 ***The old CB radio channels*** I became aware of this fact through the Wikipedia entry for "Online disinhibition effect," which cites Kenneth Tynan, "Fifteen Years of the Salto Mortale," *The New Yorker*, February 20, 1978, as the original source.

158 ***the Jerky Boys*** For anyone interested in the world of phone-call humor, Longmont Potion Castle is the Mitch Hedberg to the Jerky Boys' Dane Cook. I could never recommend the *Longmont Potion Castle II* album strongly enough.

158 ***People still flame one another*** See Todd Dugdale, "Sandbaggers and Trolls," kd0tls Ham Radio Experience, January 6, 2014, kd0tls.blogspot.com/2014/01/sandbaggers-and-trolls.html.

158 ***The government has the greatest vested*** My discussion of government surveillance of unrest, and the work of Peter Gloor at MIT, draws from "What Makes Heroic Strife," *Economist*, April 21, 2012, economist.com/node/21553006.

160 ***27.5 percent of Twitter's 500 million tweets*** This number is from analysis of my randomized research sample.

160 ***Facebook's data team*** Facebook's data analysis is always done with anonymized and aggregated data. This discussion of iterations surrounding the "No one should . . ." meme, and the attendant table, was drawn from Lada Adamic et al., "The Evolution of Memes on Facebook," January 18, 2014, facebook.com/notes/facebook-data-science/the-evolution-of-memes-on-facebook/10151988334203859. The post leaves it unclear how political bias was determined. My best guess is from users' "like" patterns.

161 ***In 1950*** This paragraph discussing polarization in American politics is based on Jill Lepore, "Long Division," *The New Yorker*, December 2, 2013.

162 ***"It has always been a mystery"*** I read *Life of Mahatma Gandhi* by Louis Fisher (New York: Harper & Brothers, 1950) in 2007, and this quote has stuck with me since.

Chapter 10: Tall for an Asian

171 ***To find out what's actually* special *to a particular group*** The method for reducing a group's collected profile text to the idiosyncratic es-

sentials I present in this chapter is my own. However, the OkCupid blog post that inspired this work—"The Real Stuff White People Like"—used a different method, developed with help from Max Shron and Aditya Mukerjee. I would not have developed my own method in this book without their prior example for that post. I developed my own method because the one used for that post had me sorting the nonsense from the "real data" as the final step. For this book, I wanted something completely algorithmic, where no human selection came into play. The method is as described—you plot the words and phrases on the grid by their percentiles and then rank them by their Euclidean distance from the desired corner of the square.

The human element came into play only in the few cases where redundant phrases, such as "my blue eyes and," "blue eyes and," and "my blue eyes" appeared on the list together. In those cases, I took the most representative word or phrase and deleted the others. The lists were not meaningfully altered by this. The method considered all phrases of four words or fewer that appeared in more than thirty profiles.

Because of space considerations three lengthy entries were pared down to avoid line wrapping. In the male antithesis table I used "follow me" instead of "follow me on instagram." In the female antithesis, I used "malcolm x" instead of "biography of malcolm x," and in the words by orientation table in the next chapter I used "feminine women" instead of "attracted to feminine women."

172 ***something called Zipf's law*** I was familiar with power law distributions already. However, I used the "Zipf's law" Wikipedia page for more information on the law. "Zipf's Law and Vocabulary," by C. Joseph Sorell, *The Encyclopedia of Applied Linguistics*, November 5, 2012, was also a resource. The table in the text was excerpted from a longer table presented in that paper.

182 ***The Irish and eastern Europeans*** From Nell Irvin Painter's *The History of White People* (New York: W. W. Norton, 2010).

182 ***in Mexico*** I lived in Mexico for several years as a child and have retained an interest in its politics. See Ronald Loewe, *Maya or Mestizo?: Nationalism, Modernity, and Its Discontents* (Toronto: University of Toronto Press, 2010).

184 ***"From empathy and sexuality"*** See Bobbi J. Carothers and Harry T. Reis, "Men and Women Are from Earth: Examining the Latent Structure of Gender," *Journal of Personality and Social Psychology* 104, no. 2 (2013): 385–407. "Men Are from ~~Mars~~ Earth, Women Are from ~~Venus~~ Earth" is the title of the article's précis: sciencedaily.com/releases/2013/02/130204094518.htm.

184 ***Aristotle looked to the emptiness*** I was already familiar with the heavens' role in Einstein's and Newton's work. For the third, older, example, I hunted around Wikipedia until I found an example I liked. See the entry for "Aether (classical element)."

Chapter 11: Ever Fallen in Love?

187 ***A few years ago a couple of MIT students*** Here, I used "Project 'Gaydar,' " by Carolyn Y. Johnson, *Boston Globe*, September 20, 2009, and the students' original paper, "Gaydar: Facebook Friendships Expose Sexual Orientation" by Carter Jernigan and Behram F. T. Mistree, *First Monday* 14, no. 10 (2009), firstmonday.org/article/view/2611/2302.

187 **The Kinsey Report *in 1948*** See Wikipedia's "Kinsey Reports" entry, which summarizes the male and female editions of Kinsey's work. The 10 percent number for men is straightforward. There is less certainty in the report around women's sexuality. The report says 2 to 6 percent of females aged twenty to thirty-five are "exclusively" homosexual.

187 ***Later studies*** See Wikipedia's "Demographics of sexual orientation" for all kinds of numbers. Also see "LGBT demographics of the United States."

187 ***"This work can usefully"*** Dan Black et al., "Demographics of the Gay and Lesbian Population in the United States: Evidence from

Available Systematic Data Sources," *Demography* 37, no. 2 (2000): 139–54.

188 ***This surely involves a painful choice*** See Assi Azar, "Op-ed: To You There, in the Closet," *The Advocate*, April 16, 2013, advocate.com/commentary/2013/04/16/op-ed-you-there-closet.

188 ***no more unusual than naturally blond hair*** My source is Professor C. George Boeree, of Shippensburg University. See his post "Race" at web space.ship.edu/cgboer/race.html. Even back-of-the-envelope math proves his point: there are roughly 1 billion Europeans, Canadians, Americans, and Australians on Earth. If 1 in 6 of them is naturally blond, which in my personal circle would be a vast overestimate, that's 2 percent of the world right there.

188 ***According to Stephens-Davidowitz*** My four-page discussion of gay porn searches and their implications adapts findings from Stephens-Davidowitz's piece "How Many American Men Are Gay?" *New York Times*, December 7, 2013. Both the Google Trends data I cite and its extension to Nate Silver's findings and to Gallup's state-by-state numbers are based on that article. Silver's original piece is "How Opinion on Same-Sex Marriage Is Changing, and What It Means," from his *New York Times* fivethirtyeight blog, fivethirtyeight.blogs.nytimes.com/2013/03/26/how-opinion-on-same-sex-marriage-is-changing-and-what-it-means.

Gallup's numbers are from Gary J. Gates and Frank Newport, "LGBT Percentage Highest in D.C., Lowest in North Dakota," gallup.com/poll/160517/lgbt-percentage-highest-lowest-north-dakota.aspx.

190 ***so does mobility data from Facebook*** In his article, Stephens-Davidowitz also extended his research into publicly available Facebook profile data.

190 ***often attributed to Thoreau*** The quote itself is a combination of a passage in Thoreau's Walden with two lines of Oliver Wendell Holmes's poem "The Voiceless." See The Walden Woods Project:

walden.org/Library/Quotations/The_Henry_D._Thoreau_Mis-Quotation_Page.

191 ***The old economic "misery index" is*** See Wikipedia's "Misery index (economics)." Arthur Okun suggested the original formulation.

196 ***"Respondents who identified"*** See Mackey Friedman, "Considerable Gender, Racial and Sexuality Differences Exist in Attitudes Toward Bisexuality," *ScienceDaily*, November 5, 2013, sciencedaily.com/releases/2013/11/131105081521.htm.

196 ***Gerulf Rieger of the University of Essex*** I reference a pair of papers by Professor Rieger and his team: Gerulf Rieger, Meredith L. Chivers, and J. Michael Bailey, "Sexual Arousal Patterns of Bisexual Men," *Psychological Science* 16, no. 8 (2005): 579–84, and its successor, Gerulf Reiger et al., "Male Bisexual Arousal: A Matter of Curiosity?," *Biological Psychology* 94, no. 3 (2013): 479–89.

197 ***Ellyn Ruthstrom*** See David Tuller, "No Surprise for Bisexual Men: Report Indicates They Exist," *New York Times*, August 22, 2011, and Meredith Melnick, "Scientific Study Finds That Bisexuality Really Exists," *Time*, August 23, 2011, healthland.time.com/2011/08/23/scientific-study-finds-that-bisexuality-really-exists.

198 ***On Facebook 58 percent*** See Chris Taylor, "Fake Facebook Users Likely to Be Popular Bisexual College Women," *Mashable*, February 3, 2012, mashable.com/2012/02/03/fake-facebook-users-bisexual-college-women.

199 ***Though people have been gay forever*** See Wikipedia's "Timeline of LGBT history" and "Coming out" entries. The idea of self-disclosure (that is, coming out) as an act of empowerment was originated by Karl Heinrich Ulrichs.

Chapter 12: Know Your Place

204 ***The United States and the USSR split Korea*** I was generally familiar with this process, mostly from *American Caesar*, but this incredible anecdote is mentioned on the "Division of Korea" Wikipedia entry, which

cites Don Oberdorfer's book *The Two Koreas* (New York: Basic Books, 2001) as the original source. I confirmed the anecdote via a search on the book's text on Google Books: books.google.com/books/about/The_Two_Koreas.html?id=yJZKpYXh2SAC.

205 ***Here you see a plot*** This map, like all the full US maps in this chapter, and the Reddit plot, was made by James Dowdell. This one was made using a standard Voronoi partition of the United States, which each Craigslist market serving as the "capital" of a "state" (called "seeds" and "cells"). Though the plot looks complex, it's actually very elegant: the segments are all the points equidistant to the two nearest seeds. I've seen various other versions of this same plot. My version was inspired by one made by IDV Solutions and posted by "john.nelson" on their UX blog: uxblog.idvsolutions.com/2011/07/chalkboard-maps-united-states-of.html.

206 ***venue of longing is Walmart*** This is the same Voronoi plot, but combined with the by-state data from Dorothy Gambrell's "Missed Connections" map, published in *Psychology Today*. The cells are coded by the top missed-connection result for the state where their seed lies. You can see the original map here: psychologytoday.com/blog/brainstorm/201302/missed-connections-0.

I transported the data to the previous Voronoi partition in order to maintain consistency with the previous Craigslist map.

206 ***Years ago, an enterprising hacker*** The hacker is Pete Warden, and his post is "How to Split Up the US," which you can find here: petewarden.com/201%2/06/how-to-split-up-the-us. As Warden notes in a later post, "Why You Should Never Trust a Data Scientist," his grouping of the United States into the seven new zones is arbitrary—the data science version of "for entertainment purposes only." I reference them here in that spirit.

207 ***Matthew Zook, a geographer*** Professor Zook and his team maintain a fantastic geography blog called Floating Sheep, and that blog was my primary source for his work: floatingsheep.org.

The earthquake discussion and the map are drawn from "Mapping the Eastern Kentucky Earthquake" posted on the Floating Sheep blog by Taylor Shelton. My image is a reproduction of the original, simplified for print: floatingsheep.org/2012/11/mapping-eastern-kentucky-earthquake.html.

The DOLLY team is Matthew Zook, Mark Graham, Taylor Shelton, Monica Stephens, and Ate Poorthuis. Poorthuis narrates the Sint Maarten walkthrough, which can be found here: www.youtube.com/watch?v=pD9HWAaQGUA.

My discussion of the student riot is drawn from the paper "Beyond the Geotag: Situating 'Big Data' and Leveraging the Potential of the Geoweb," by Jeremy W. Crampton et al., *Cartography and Geographic Information Science* 40, no. 2 (2013): 130–39.

210 ***Below is a plot of gay porn downloads*** IP address does not pinpoint any one person (or, more precisely, a computer address) to their exact location, only to a range of about ten to fifty miles. It is roughly the same technology used by, say, weather.com, to guess at what city's weather to show you by default before you tell it a zip code. It only tells the general area from which a computer is accessing the Internet. From this research, we know nothing about the computers themselves other than what porn they were downloading; and we know absolutely nothing about who was actually using the computer, or in some cases, if there was even a person involved at all.

213 ***a forty-year-old woman in the Bay Area*** See "I'm Just Gonna Throw This Out There. Any Redditors in the SF Bay Area Have a Empty Spot at Their Table for a Lonely Thanksgiving Orphan?" posted by user "MeMyselfOhMy" on Reddit: reddit.com/r/AskReddit/comments/ebhh1.

213 ***topics that you'll only find on Reddit*** The example posts mentioned were all on the front page of their respective subreddits on January 30, 2013.

216 ***Anderson's main topics are nationalism*** Showing the flexibility of his theory, many of Anderson's ideas on nationhood are surprisingly applicable to online communities. He describes nations as "both inherently limited and sovereign" and "conceived as a deep, horizontal comradeship." And especially applicable to the Internet is this passage: "This new synchronic novelty could arise historically only when substantial groups of people were in a position to think of themselves as living lives *parallel* to those of other substantial groups of people—if never meeting, yet certainly proceeding along the same trajectory." Benedict Anderson, *Imagined Communities* (London: Verso, 1983), 6, 191–92.

217 ***a worldwide look at modern large-scale movements*** I obtained permission from the Facebook researchers Aude Hofleitner, Ta Virot Chiraphadhanakul, and Bogdan State to reproduce their map and discuss their results. They asked that I include a more robust explanation of "coordinated migration" and of their study. Here are their words:

> In a coordinated migration, a significant proportion of the population of a city has migrated, as a group, to a different city. More specifically, a flow of population from city A (hometown) to another city B (current city) is considered a coordinated migration if, among the cities in which people from hometown A currently live, city B is the city with the largest number of individuals with current city B, and hometown A. There are numerous migrations to, from, and within the United States but they do not exhibit this coordinated property because there is no overly dominant attractive city and people move to different areas. This map displays chunks of the small towns and villages of Southeast Asia relocating en masse, in a coordinated fashion, to the urban centers.

For more information and the full study, please refer to the Facebook Data Science post on Coordinated Migration:

www.facebook.com/notes/facebook-data-science/coordinated-migration/10151930946453859.

As you'll see when you visit the link, in reproducing their work, I modified their original map by removing the labels and focusing on a smaller part of the region, to make the map more readable in print. Thank you to Mike Develin, also at Facebook, for helping facilitate permission for this reproduction. All Facebook Data Science work is done on anonymized and aggregated data.

Chapter 13: Our Brand Could Be Your Life

221 ***But what they don't tell you*** See Clare Baker, "Behind the Red Triangle: The Bass Pale Ale Brand and Logo" Logoworks.com, November 8, 2013, logoworks.com/blog/bass-pale-ale-brand-and-logo.

221 ***Archaeologists have unearthed*** My discussion of branding in ancient times is based on David Wengrow, "Prehistories of Commodity Branding," *Current Anthropology* 49, no. 1 (2008): 7–34, and Gary Richardson, "Brand Names Before the Industrial Revolution," NBER Working Paper No. 13930, National Bureau of Economic Research, Cambridge, MA, 2008. http://papers.nber.org/paper/w10411.

221 ***In 1997, Tom Peters*** See "The Brand Called You" by Tom Peters, published in *Fast Company*, August/September 1997, fastcompany.com/28905/brand-called-you.

222 ***still read in marketing classes*** See "What a great article. I was given this to read for a class of mine, and it is written brilliantly. Great insight and information on branding. Thanks!!" a comment by user "Morgan" on Peter's article on Fastcompany.com.

222 ***a man named Peter Montoya*** Montoya's first work on the topic was titled *The Brand Called You: The Ultimate Brand-Building and Business Development Handbook to Transform Anyone into an Indispensable Personal Brand*, by Peter Montoya and Tim Vandehey (self-published, 2003). This was then republished as *The Brand Called You: Make Your Business Stand Out in a Crowded Marketplace*

(New York: McGraw-Hill, 2008), which according to Amazon was an "international bestseller." A PDF of the first chapter is hosted here: petermontoya.com/pdfs/tbcy-chapter1.pdf. Montoya's personal site redirects to marketinglibrary.net, where you can book him for speaking engagements.

223 ***You can see the birth of the idea*** For this chart, I subtracted the long standing idiom of "personal brand of" (as in "personal brand of leadership") from the results for "personal brand" to isolate the self-marketing phenomenon.

223 ***Dale Carnegie*** I relied on Wikipedia's "Dale Carnegie" entry for basic details on his life.

224 ***For every kid who tweets herself*** The two incidents I allude to here are Bernie Zak's campaign to get into UCLA, as detailed in Brock Parker, "Brookline Student Lobbies UCLA on Twitter" *Boston Globe*, May 7, 2013, and Rob Meyer's hiring by the *Atlantic Monthly*, as described in Alexis C. Madrigal, "How to Actually Get a Job on Twitter," *Atlantic Monthly*, July 31, 2013.

See also Jason Fagone, "The Construction of a Twitter Aesthetic," *The New Yorker*, February 12, 2014, newyorker.com/online/blogs/culture/2014/02/the-construction-of-a-twitter-aesthetic.html.

226 ***the different way African Americans tend*** My discussion of Black Twitter drew on the following sources:

Choire Sicha, "What Were Black People Talking About on Twitter Last Night?" *The Awl*, November 11, 2009, theawl.com/2009/11/what-were-black-people-talking-about-on-twitter-last-night.

Farhad Manjoo, "How Black People Use Twitter," *Slate*, August 10, 2010, slate.com/articles/technology/technology/201%8/how_black_people_use_twitter.html. A counterpoint to Manjoo's piece is "Why 'They' Don't Understand What Black People Do on Twitter" by Dr. Goddess, on blogspot. Goddess especially objects to the portrayal of blacks on Twitter as a "monolith"—the word appears twice in the post, and I echo it in my discussion. See drgoddess.blogspot.com/201%8/why-they-dont-understand-what-black.html.

"How to Be Black Online," a slideshow by Baratunde Thurston, is a clever overview of Black Twitter and acknowledges better than most sources that, like many racial tropes, "Black Twitter" is both "funny because it's true" and inaccurate at the same time. See slideshare.net/baratunde/how-to-be-black-online-by-baratunde.

Hard data on Twitter usage by ethnicity can be found in the Pew Research report "Demographics of Key Social Networking Platforms" (2013), by Maeve Duggan and Aaron Smith: pewinternet.org/2013/12/30/demographics-of-key-social-networking-platforms.

For evidence of white confusion over Black Twitter, see Nick Douglas, "Micah's 'Black People on Twitter' Theory," Too Much Nick, August 21, 2009, toomuchnick.com/post/168222309.

227 ***Right now there are 2,643*** The site Social Bakers ranks all Twitter accounts by number of followers. The number has, no doubt, changed. Visit socialbakers.com/twitter/ and page back through the rankings to see for yourself. For information on US taxpayers by income, visit the IRS's "SOI Tax Stats—Individual Statistical Tables by Filing Status" page at irs.gov/uac/SOI-Tax-Stats---Individual-Statistical-Tables-by-Filing-Status. Information on the Forbes Billionaires list is from Elizabeth Barber, "Forbes' Richest People: Number of Billionaires up Significantly," *Christian Science Monitor*, March 3, 2014, csmonitor.com/USA/USA-Update/2014/0303/Forbes-richest-people-number-of-billionaires-up-significantly-video.

227 ***Newt Gingrich boasted*** See Jeff Neumann, "Newt Gingrich Brags About His Twitter Followers," *Gawker*, August 1, 2011, gawker.com/5826477. Also see John Cook, "Update: Only 92% of Newt Gingrich's Twitter Followers Are Fake," *Gawker*, August 2, 2011, gawker.com/5826960.

227 ***Mitt Romney*** See "Is Mitt Romney Buying Twitter Followers?" by Zach Green on 140elect: 140elect.com/twitter-politics/is-mitt-romney-buying-twitter-followers. My data and chart are adapted from the data and chart in that post.

229 ***"We, the users"*** See Jenna Wortham, "Valley of the Blahs: How Justin Bieber's Troubles Exposed Twitter's Achilles' Heel," *New York Times* Bits blog, January 25, 2014, bits.blogs.nytimes.com/2014/01/25/valley-of-the-blahs-how-justin-biebers-downfall-exposed-twitters-achilles-heel.

230 ***In 2012, Salesforce.com*** My discussion of Salesforce's job post draws on the following sources:

Drew Olanoff, "Klout Would Like Potential Employers to Consider Your Score Before Hiring You. And That's Stupid," TechCrunch, September 29, 2012, techcrunch.com/2012/09/29/klout-would-like-potential-employers-to-consider-your-score-before-hiring-you-and-thats-stupid.

Jessica Roy, "Want to Work at Salesforce? Better Have a Klout Score of 35 or Higher," *BetaBeat*, September 27, 2012, betabeat.com/2012/09/you-may-not-work-at-salesforce-unless-you-have-a-klout-score-of-35-or-higher.

The original job posting was still active when I was writing, but has since been removed.

231 ***The gates open and close*** See Larry Wissel, "How Does a Logic Gate in a Microchip Work? A Gate Seems Like a Device That Must Swing Open and Closed, Yet Microchips Are Etched onto Silicon Wafers That Have No Moving Parts. So How Can the Gate Open and Close?" *Scientific American*, "Ask the Experts," October 21, 1999, scientificamerican.com/article/how-does-a-logic-gate-in.

The gates on a microchip aren't doors in the traditional sense, swinging on tiny hinges. They use voltage to control movement, whereas an old gate might use wooden slats. But they, like gates, control flow from one space to another, and are either open or shut.

233 ***Target, by analyzing a customer's purchases*** See Kashmir Hill, "How Target Figured Out a Teen Girl Was Pregnant Before Her Father Did," *Forbes*, February 16, 2012, forbes.com/sites/kashmirhill/2012/02/16/how-target-figured-out-a-teen-girl-was-pregnant-before-her-father-did.

233 ***a Jell-O marketing campaign*** The Jell-O discussion and illustrative tweets are drawn from Harry Bradford, "Jell-O's Fun My Life Twitter Campaign: Social Media Genius or Just 'Funning' Annoying?" *Huffington Post*, May 24, 2013, huffingtonpost.com/2013/05/24/jello-fun-my-life-twitter_n_3332230.html.

233 ***McDonald's sent out*** Drawn from Hannah Roberts, "#McFail! McDonalds' Twitter Promotion Backfires as Users Hijack #Mcdstories Hashtag to Share Fast Food Horror Stories," *Daily Mail*, January 24, 2012, dailymail.co.uk/news/article-2090862.

234 ***Wendy's had tried*** Drawn from "When Twitter Hashtag Promotion Marketing Goes Bad #HeresTheBeef" by blogger "stacie," on the Divine Miss Mommy blog: thedivinemissmommy.com/when-twitter-hashtag-promotion-marketing-goes-bad-heresthebeef.

234 ***More recently, Mountain Dew*** See Everett Rosenfeld, "Mountain Dew's 'Dub the Dew' Online Poll Goes Horribly Wrong," *Time*, August 14, 2012, newsfeed.time.com/2012/08/14/mountain-dews-dub-the-dew-online-poll-goes-horribly-wrong.

Chapter 14: Breadcrumbs

239 ***As of May 2013, Facebook was recording*** See Craig Smith, "By the Numbers: 98 Amazing Facebook Stats," Digital Marketing Ramblings, March 13, 2014, expandedramblings.com/index.php/by-the-numbers-17-amazing-facebook-stats/#.U1AArPldXko.

239 ***a group from the UK*** This passage and the table are based on "Private Traits and Attributes Are Predictable from Digital Records of Human Behavior," by Michal Kosinskia, David Stillwell, and Thore

Graepel, *Proceedings of the National Academy of Sciences* 110, no. 15 (2013): 5802–5805.

241 ***Xbox One*** See Stephen Fairclough, "Physiological Data Must Remain Confidential," *Nature* 505, no. 7483 (2014): 263.

241 ***The UK has 5.9 million*** See David Barrett, "One Surveillance Camera for Every 11 People in Britain, Says CCTV Survey," *Telegraph*, July 10, 2013, telegraph.co.uk/technology/10172298.

241 ***In Manhattan*** See Brian Palmer, "Big Apple Is Watching You," *Slate*, May 3, 2010, slate.com/articles/news_and_politics/explainer/201%5/big_apple_is_watching_you.html.

243 ***All those security cameras*** See Jon Healey, "Surveillance Cameras and the Boston Marathon Bombing," *Los Angeles Times*, April 17, 2013, articles.latimes.com/2013/apr/17/news/la-ol-boston-bombing-surveillance-suspects-20130417.

See also "The Need for Closed Circuit Television in Mass Transit," by Michael Greenberger, University of Maryland Legal Studies Research Paper No. 2006–15, Law Enforcement Executive Forum (2006): 151, digital commons.law.umaryland.edu/cgi/viewcontentcgi?article=1065&context=fac_pubs.

244 ***"master the Internet"*** This phrase in particular refers to the NSA's cooperation with the surveillance apparatuses of other governments, as part of the "Five Eyes" Alliance. See Wikipedia's "Mastering the internet" entry. The slide depicted was widely circulated after its publication by the *Guardian*. See theguardian.com/world/interactive/2013/nov/01/prism-slides-nsa-document.

245 ***"For each of the millions"*** See David Medine et al., "Report on the Telephone Records Program Conducted under Section 215 of the USA PATRIOT Act and on the Operations of the Foreign Intelligence Surveillance Court," Privacy and Civil Liberties Oversight Board (2014), http://www.fas.org/irp/offdocs/pclob-215.pdf.

247 ***Women are using apps*** My discussion of menstruation apps is

based on Jenna Wortham, "Our Bodies, Our Apps: For the Love of Period-Trackers," *New York Times*, January 23, 2014.

247 ***there's a startup that says it can* infer** This fact is from Jaron Lanier, "How Should We Think About Privacy?" *Scientific American*, November 2013, 65–71.

247 ***all the analysis was done anonymously and in aggregate*** It bears repeating that at no time was any data tied back to any individual. For the user photos and text cited in the book see the notes above related to them.

248 ***Jaron Lanier*** My discussion of Lanier's work focuses on his article "How Should We Think About Privacy?"

248 ***"Using data drawn from queries"*** See John Markoff, "Unreported Side Effects of Drugs Are Found Using Internet Search Data, Study Finds," *New York Times*, March 7, 2013, nytimes.com/2013/03/07/science/unreported-side-effects-of-drugs-found-using-internet-data-study-finds.html.

248 ***a crowdsourced family tree*** Geni.com reports more than 75 million entries in its tree. They're owned by MyHeritage, which claims 1.5 billion.

249 ***two political scientists debunked*** See Jowei Chen and Jonathan Rodden, "Don't Blame the Maps," *New York Times*, January 26, 2014, nytimes.com/2014/01/26/opinion/sunday/its-the-geography-stupid.html.

249 ***Republicans owe their House majority*** See nytimes.com/2014/01/26/opinion/sunday/its-the-geography-stupid.html

249 ***Facebook was collecting 500 terabytes*** See Eliza Kern, "Facebook Is Collecting Your Data—500 Terabytes a Day," Gigaom, August 22, 2012, gigaom.com/2012/08/22/facebook-is-collecting-your-data-500-terabytes-a-day.

249 ***Alex Pentland at MIT*** My discussion of Pentland draws on his article "Reality Mining of Mobile Communications: Toward a New Deal on

Data," in *Global Information Technology Report 2008–2009*, ed. Soumitra Dutta and Irene Mia (Geneva: World Economic Forum, 2009), 75–80, and an interview with him, "An Interview with Alex 'Sandy' Pentland About 'Social Physics' " by IDcubed: idcubed.org/?post_type=home_page_feature&p=880.

251 ***The* Washington Post *captures the shortfall*** See "Million Mask March descends on Washington" on the *Washington Post*'s PostTV blog: http://wapo.st/1b5Kt5J.

Coda

253 ***Tufte's books*** The discussion of the Vietnam Memorial, and the quote I use, are from *Beautiful Evidence* (Cheshire, CT: Graphics Press, 2006), but Tufte's *Envisioning Information* (Cheshire, CT: Graphics Press, 1990) and *The Visual Display of Quantitative Information* (Cheshire, CT: Graphics Press, 2001) were also indispensible.

254 ***The memorial was digitized in 2008*** See fold3.com/thewall and Mallory Simon, "Vets Pay Tribute to Fallen Comrades at Virtual Vietnam Wall," CNN.com, April 1, 2008, cnn.com/2008/TECH/04/01/vietnam.wall.

254 ***Two pictures had been added to his entry*** PFC Wilson's profile on fold3 is at fold3.com/page/631972608_lorne_john_wilson/stories. It is unclear if he is personally depicted in the group picture. It's clearly an authentic snapshot from the Vietnam War, but it is blurry.

Acknowledgments

Like pages without binding, this project and indeed my life would've flown to the winds long ago without my wife, Reshma. Thank you for your unwavering support, selflessness, and love.

Thank you to Max Krohn, Sam Yagan, and Chris Coyne for building OkCupid and for having me along. It has been a privilege to work with and for you guys for the last fifteen years.

Thank you to my agent, Chris Parris-Lamb, who turned *Dataclysm* from a rambling pitch at a bar into a bona fide proposal, and to Amanda Cook, my editor at Crown, who took it from there. To the extent this book is a success, her patience and skill have made mere ideas into something worth reading. Thank you also to Emma Berry, editorial assistant, and to the design team, especially Chris Brand, for bringing *Dataclysm* into being, and to Annsley Rosner, Sarah Breivogel, Sarah Pekdemir, and Jay Sones for helping it out into the world. The support and vision of Molly Stern, Jacob Lewis, and David Drake made all of the above possible. Thank you, too, to Allison Lorentzen at Penguin for her very early guidance into the publishing world.

Thank you to James Dowdell, my versatile data researcher and programmer. James did the essential database work behind *Dataclysm* and also generated many of the book's maps and network plots. Thank you to Tom Quisel and Mike Maxim for pulling (and repulling!) data from OkCupid, and for being excellent sounding boards for my various statistical ideas.

Thank you to my parents and my sister for their encouragement and for being the foundation of my life. Thank you to the Patel family for supporting me and, especially, Reshma, while we bent our days and weeks and months around getting this book finished.

Thank you to Eddie Lou at Shiftgig, Tim Abraham at StumbleUpon (and now Twitter), Ryan Ogle and Sean Rad at Tinder, Jim Talbot at Match, Tom Jacques at DateHookup, and Erik Martin at Reddit for aggregated data and access. Thank you to Michael Tapper and Ben Murray for reading drafts, and to Sean Mathey at Mathey & Tree, Eric Brown at Franklin, Weinrib, Rudell & Vassallo, and John Therien at Smith Anderson for legal work. Thank you to Doug Demay for advice that was no less wise for being informal. Finally, thank you to Jed McCaleb and Justin Rice, who, from d20s to bitcoin to Dylan to *Ulysses*, have taught me so much. My life and this book are much richer for your friendship.

Index

Page numbers in *italics* refer to illustrations.